Big Fleas Have Little Fleas

Big Fleas Have

Little Fleas

How Discoveries

of Invertebrate Diseases Are

Advancing Modern Science

Elizabeth W. Davidson

The University of Arizona Press / *Tucson*

The University of Arizona Press
© 2006 The Arizona Board of Regents
All rights reserved
∞ This book is printed on acid-free, archival-quality paper.
Manufactured in the United States of America

11 10 09 08 07 06   6 5 4 3 2 1

Library of Congress Cataloging-in-Publication Data
Davidson, Elizabeth W.
Big fleas have little fleas : how discoveries of invertebrate diseases are
advancing modern science / Elizabeth W. Davidson.
p. cm.
Includes bibliographical references and index.
ISBN-13: 978-0-8165-2612-3 (alk. paper)
ISBN-10: 0-8165-2612-5 (alk. paper)
ISBN-13: 978-0-8165-2544-7 (pbk. : alk. paper)
ISBN-10: 0-8165-2544-7 (pbk. : alk. paper)
1. Invertebrates—Diseases. 2. Pathogenic microorganisms. I. Title.
QR325.D38 2006
592—dc22

                                                        2006006192

To the memory of Dr. John D. Briggs
(1926–2002),
mentor, role model, and friend,
who introduced me
to the small but international world
of invertebrate pathology
and to the wonderful group of scientists
who are fascinated with these
unusual organisms.

So, Nat'ralists observe, a Flea
Hath smaller Fleas that on him prey;
and these have smaller Fleas to bite 'em;
and so proceed ad infinitum.

—*Jonathan Swift*

Contents

Illustrations

Acknowledgments

I am grateful to the many agencies and invertebrate pathology colleagues who kindly provided articles and photographs and made comments on various chapters. These include the Institut Pasteur Photographic Collection, the Ohio Agricultural Research and Development Center Library, the Society for Invertebrate Pathology Slide Collections, World Health Organization, Thomas Angus, Mark Goettel, Allan Yousten, Robert Granados, Mike Klein, Marla Spivak, Ann Hajek, Keio Aizawa, Don Roberts, Larry Gringorten, Toshi Iizuka, Jaroslav Weiser, Alois Huger, Colin Berry, Randy Gaugler, Brian Federici, and Donald Lightner. Several undergraduate students—including Anna McClaugherty, Becky Sears, Meredith Boley, and Luke Parr—also read and made suggestions on chapters. Mia McNulty performed important research for chapters 1 and 15. Some of the photographs used in this book are from the collection of the late John D. Briggs, and I am grateful to his daughter, Laura Briggs Gordon, for donating her father's large photograph collection to me.

I extend special thanks to John Alcock and David Brown, who read the manuscript several times and made very useful comments. The sabbatical that I spent at Mount Union College (in Alliance, Ohio) in 2002 was invaluable to the writing of much of this manuscript, and I am very grateful to the Department of Biology at Mount Union College for their collegiality and for the opportunity to concentrate on the manuscript. Allyson Carter of the University of Arizona Press provided valuable assistance, encouragement, and enthusiasm during the final stages of the manuscript preparation.

Above all, I am grateful to my husband, Joseph Davidson, for his patience and understanding during the more than ten years that this manuscript has been in progress.

Big Fleas Have Little Fleas

1

Pasteur, Silkworms, and the Germ Theory of Disease

Sooner or later, every living thing gets sick. Viruses, fungi, and bacteria infect plants, butterflies, and crabs just as they do us. But we also depend on microorganisms to recycle nutrients in the soil, make cheese, raise bread, ferment wine, and produce antibiotics to cure our diseases. Similarly, invertebrate animals rely on microorganisms for nutrition and health. Although these interactions have been going on for as long as there have been living things on earth, in recent decades science has found some amazing uses for microorganisms associated with invertebrates. The stories told here highlight a few of these associations that have become particularly important to humans.

But these are also stories of scientists who became fascinated with strange microbes living with invertebrates. Those scientists learned to use diseases to control pest populations and to make products that have already provided great benefits to us all and hold promise for even greater benefits in the future. And some of those scientists have learned ways to detect and prevent disease in the most valuable invertebrates—those we rely on for our food supply.

The first diseases of lower animals to catch the eye of observant people were afflictions of the two insects that have long been regarded as "the good guys" because they are valuable to humans: honeybees and silkworms, the two insects that have been domesticated. Indeed, diseases of bees were described in myths as early as 700 BC, and the silkworm has been used to make valuable cloth for at least five thousand years.

The principal silkworm is a moth, *Bombyx mori*, which was domesticated perhaps as early as 3000 BC in China (a different species was domesticated a thousand years earlier in India). The silkworm is the only truly domestic insect, because—unlike the honeybee—it cannot survive without human care, and *Bombyx mori* no longer exists in the wild. The true source of the luxurious silk cloth, as valuable as gold, was kept a secret in China for more than two thousand years. Anyone who attempted to take a silkworm out of the country was punished by death. Nonetheless, in AD 555, two brave monks smuggled silkworm eggs concealed inside a staff from China to Constantinople, and this act led to the establishment of the European and Middle Eastern silk industries. Spain and Italy produced silk as early as the 1100s and 1200s, and from there the culture of silkworms eventually spread to France and other parts of Europe. Clell L. Metcalf and Wesley P. Flint, in their 1928 book *Destructive and Useful Insects*, estimated that the annual value of silk production was then $200–500 million; now it is many times that figure. Thanks to efficient modern rearing techniques of silkworms in China, Japan, and India, silk clothing is now both popular and affordable.

Adult silkworm moths are creamy white and about two inches long (fifty millimeters). Their function is simply to reproduce; they fly very weakly if at all, and they never eat. The female lays three hundred to four hundred eggs during her two- or three-day adult life. Larvae that hatch from these eggs grow over the next three to four weeks into fat caterpillars. Silkworm caterpillars are reared in large baskets in long sheds, where they are fed fresh mulberry leaves daily. At the end of the caterpillar stage, they spend about three days spinning a silken cocoon and forming the pupa (or chrysalis) inside. The adult moth will emerge from this cocoon two to three weeks later. Most of the pupae, however, do not survive to adulthood, as the cocoons are dropped into hot water or exposed to chemicals to kill the insect inside. The cocoons are then soaked in warm water to loosen the threads. The silkworm has a unique characteristic: it is the only insect that produces a single unbroken thread of silk that can be unrolled from the cocoon and gathered on a spindle, ready for weaving into cloth.

When many thousands of silkworms are reared together, the stage is set for infectious diseases. "Stiffened" caterpillars, which died from a fungus disease known as muscardine, were used in early Chinese medicine. As silk culture became popular in Europe, growers recognized several different caterpillar diseases besides the muscardine fungus disease: pebrine, a speckled condition resembling pepper grains on the skin of the caterpillar; grasserie, in which the larvae became soft and melted; and flacherie, which turned the larvae black. These diseases caused such damage to the European silk industry that they caught the eye of two brilliant scientists, and as a result an insect larva became a player in the discovery that microorganisms can cause disease.

Around 1800, silk production in Italy was plagued by a white fungus disease called calcino. At that time, diseases were widely regarded as the result of "bad air," and maggots, worms, and fungi were thought to arise by spontaneous generation. Microorganisms had been observed by the Dutch botanist Antoni van Leeuwenhoek in 1680, but the connection between microorganisms and disease had not been confirmed. Then a brilliant young Italian, trained in both law and the sciences, became fascinated with this calcino disease. Agostino Maria Bassi of Lodi received an inheritance that permitted him to give up law and focus on how this disease was generated—by spontaneous generation, he imagined at the beginning. He spent five years, from 1808 to 1813, trying to prove that this theory was correct. After finally rejecting spontaneous generation, he set up a series of scientifically sound experiments exploring the hypothesis that a "germ" was somehow involved.

Agostino Maria Bassi of Lodi, Italy

Through many years of research, he eventually showed that the calcino disease was caused by a "vegetable parasite" and that "seeds" from the sick insects could be transferred from infected to healthy silkworms. He correctly observed that this disease is caused by a fungus (later named *Beauveria bassiana* in his honor) that killed the silkworms in warm, humid conditions. He also showed that the disease could pass to caterpillars other than the silkworm, and he came up with techniques to prevent the disease in silkworm culture. Finally, in 1834 Bassi presented the results of his twenty-five years of work to the Faculties of Medicine and Philosophy at the University of Padua. He continued to work into his late seventies, studying tuberculosis, cholera, and pellagra. Because Bassi was the first person to accurately make the connection between a microorganism and a disease, he is considered by many historians to be the father of the "germ theory" of disease.

Beginning in 1849, the highly profitable French silk industry suffered from a disease outbreak of such proportions that production fell to one-sixth its original value. Losses equaled hundreds of millions of francs. Even the importation of silkworm eggs at great expense from supposedly disease-free areas in Japan did not relieve the problem. By 1865 the disease had spread to other areas of Europe, as well as Asia and the Middle East. A large group of silk growers petitioned the French senate for assistance, which resulted in the enlistment of Louis Pasteur to investigate the problem.

As a chemist, Pasteur had earlier studied the right- and left-handed forms of some molecules, and at the time was studying fermentation and "diseases" of wine. These studies, which he had begun in 1854, eventually led to a germ theory of fermentation. Wine is the result of the activity of yeast on grape juice, resulting in the production of alcohol. However, wine can also become vinegar (*vin aigre*, or "sour wine," in French) if certain bacteria are also present, that is, when the wine becomes "diseased." Later, Pasteur demonstrated that the application of heat before shipping preserved the flavor and quality of the wine. He stated prophetically, "There could be a tremendous potential market for French wines in foreign countries." This technique, called pasteurization, is used to this day to preserve milk, fruit juices, and other food products.

Pasteur initially turned down the invitation to study silkworm

diseases, as he knew nothing about insects. However, his mentor and
friend Jean-Baptiste Dumas (senator from the Le Gard district of
southern France, where the disease was causing devastation to silk
production) urged him to accept the assignment. Out of respect for
Dumas, Pasteur reluctantly accepted the challenge. And so, in 1865,
Louis Pasteur and his family traveled to the south of France, to the
village of Alés in the remote and mountainous Cévennes region, to
identify the cause of the silkworm diseases.

Shortly after arriving in Alés, Pasteur sought out the famous scientist J. Henri Fabre, founder of the science of entomology, to learn
about the life cycle of the silkworm. Fabre showed him a cocoon.
Upon shaking the cocoon, Pasteur was surprised to find that it rat-
tled. When he asked Fabre whether something was inside and was
told that it contained a chrysalis, he did not know what this could be.
Fabre explained that it was a sort of mummy in which the caterpillar
transformed into the moth. For this reason the caterpillar spun the
valuable silk, to protect itself during this transformation. Pasteur's
only response to this was "Ah!"

The first disease to command Pasteur's attention was pebrine,
which caused tiny black spots to develop on the surface of the caterpillars. Pasteur talked to workers about their experiences with this
disease, and he found a small silkworm nursery where he could carry
out his experiments. Being a chemist, he did not immediately succeed in discovering the cause of this disease; indeed, for almost two
years, he was convinced that it was metabolic or chemical in nature.
Finally, however, he and his assistants became convinced that the
disease was caused by tiny "corpuscles"—microscopic oval objects
that they could see in the blood of infected silkworms.

Finding the cause of this disease was complicated by the observation that offspring of an infected female moth were infected regardless of how clean they were kept, but offspring of clean mothers
were infected by contact with sick larvae. Finally Pasteur realized
that the corpuscles were being transmitted both between caterpillars and from the mother to the offspring through the egg. It is a
tribute to Pasteur's insight that he understood that this agent could
be transmitted in more than one way. The cause of pebrine is now

Louis Pasteur and his wife at Alés, where he studied silkworm diseases

known to be a protozoan, *Nosema bombycis*, a one-celled parasite with
a complex life cycle including a resistant spore that can be transmitted either from the mother to the young, or by contamination of the larval diet by the feces of infected insects.

Pasteur devised a simple method of ensuring that the eggs used for the next generation of silkworms were free of pebrine disease. After the female moth had laid her eggs on a bit of paper, she was examined for the presence of corpuscles, and if these were found, she and her eggs were burned. If no corpuscles were seen, the eggs were kept and hatched. This technique was soon adopted in France, Austria, Italy, and Asia and saved the silk industry in these regions.

At first, silkworm farmers were skeptical about the need to use the "complicated" microscope to detect this disease. The compound microscope had been developed only about thirty years earlier, and the farmers did not know how to use this new instrument. But Pasteur pointed out that even his eight-year-old daughter, Marie-Louise, had learned to use it without difficulty. Microscopes soon became widely used in silk production. Because the ability to diagnose disease would undercut the profitable importation of silkworm eggs, however, importers fought the technique and published attacks against Pasteur in trade journals. Nonetheless, this technique remains a simple and efficient method of ensuring the health of the next generation of silkworms and is currently required by law in Japan.

Just when Pasteur had solved the pebrine mystery, he found dead caterpillars that were free of corpuscles. He realized that there must be two diseases, and he established that the second disease, called flacherie, was caused by bacteria. The bacteria appeared to arrive on leaves, and when the caterpillars were in poor housing conditions and had weakened resistance, they succumbed to bacterial disease. Growers soon began selective breeding for a stronger strain of silkworms.

Pasteur wrote a two-volume memoir, *Études sur la maladie des vers a soie* (*Studies of Silkworm Disease*), which was published in 1870. Many important ideas came to Pasteur during his work on the silkworm, including the novel suggestion that susceptibility to disease could be either hereditary or acquired. He speculated on the origins of epidemics and was the first to use the term *biological chemistry*.

Pasteur's research in Alés was interrupted by several personal tragedies, including the deaths of his father and two daughters, aged two and twelve years. In 1868, Pasteur suffered a cerebral hemorrhage, which left him permanently with little use of his left hand and leg. Nonetheless, he continued his work, assisted by his wife and their daughter Marie-Louise.

In 1867 Pasteur had requested that the government build a new laboratory in Paris for the study of infectious diseases. The emperor, Napoleon III, immediately approved the request. Work on the laboratory began, but the Franco-Prussian War delayed its completion. After Pasteur's stroke, work on the laboratory was suspended. Finally, however, Pasteur convinced the emperor and the government that he was returning to work, and the laboratory was completed.

Pasteur returned to Paris to work in his new laboratory, the Pasteur Institute. There he studied anthrax, rabies, staphylococcus, streptococcus, septicemia, and other infectious diseases of humans and domestic animals. His research—along with that of contemporary scientists such as Robert Koch—provided the final blow to the theory of spontaneous generation. From his laboratory at the Pasteur Institute came early virus vaccines. The Pasteur Institute continues to produce world-class science to the present day. Indeed, critical discoveries made at the institute, such as the virus causing AIDS, have led to several Nobel Prizes. In 1854 Pasteur observed, *"Dans les champs de l'observation le hasard ne favorise que les esprits preparés"* ("In the field of observation, chance favors only the prepared mind"). Pasteur's mind was prepared for his great work on human disease by insights that he gained from the diseases of the lowly silkworm.

2

Of Caterpillars and Crystals

Bacterial Toxins as Insecticides

As you drive through the American countryside on a summer afternoon, you will pass acre after acre of soybeans, corn, and (in the South and Southwest) cotton, three of the country's most valuable crops. But what you may not know is that almost half of those plants are protected from insects by genes from one insecticidal bacterium, *Bacillus thuringiensis*. The discovery and development of *Bacillus thuringiensis* as an important insect control agent was the result of scientists working independently in several corners of the world for more than a century.

In Japan in the late 1800s, drably clad young women in bare feet, with hair bound up in cloth and wearing simple kimonos, were carefully feeding the valuable silkworm caterpillars. It was a hot and sweaty job. Each adult female moth was tethered by a thread to a piece of paper, where she laid her eggs. Once the eggs hatched, the women brushed the caterpillars into large shallow baskets and fed them large quantities of finely chopped mulberry leaves. Finally, when the caterpillars were about to pupate and produce their silken cocoons, the young women used feathers to gently brush them into large trays. Later the cocoons were heated in pans of water to kill the insect and release the silk thread, which was wound off each cocoon onto a wheel-like device, stretched over wooden posts, and finally hung in skeins over rods. This thread was spun, dyed, and woven into elegant, valuable silk cloth.

Thousands of caterpillars confined in a hot, humid environment, fed a daily diet of fresh mulberry leaves from the surrounding countryside, and attended by women moving from one tray to the next set the stage for disease outbreaks. One of these diseases was called sudden death syndrome.

In 1898, a Japanese scientist named Shigetane Ishiwata began to study this sudden death syndrome. Using a microscope, Ishiwata saw a rod-shaped bacterium with an oval spore. This bacterium killed caterpillars with amazing speed. If a caterpillar swallowed some of these bacteria, it stopped eating almost immediately. It held up its head, lost control of its legs, fell over, stretched out its body, and within minutes was dead. A few hours later, the body turned brown and then black and dissolved into a liquid—a ghastly death indeed! Ishiwata named the bacterium *Bacillus* ("rod") *sotto* ("sudden collapse"). He called the disease *sotto-byo-kin* ("collapse-disease-microorganism"), and it became known as sotto-kin.

Ishiwata was a keen observer and practiced the scientific method just as we do today. He found that cultures grown on agar were most active after one week of growth but remained lethal for eight or nine months and that spores could survive up to seven years. In addition, he observed that caterpillars died before the bacteria grew within them. Ishiwata concluded that they were killed by a toxin that was "closely on or in the spores." Although colleagues who examined the bacterium felt it might be one that also killed bees, Ishiwata argued that it was a brand new species of bacillus associated only with moths and butterflies.

The discovery of sotto-kin led to a flood of experiments on *Bacillus sotto* in Japan from its first discovery in 1901 until the mid 1930s. K. Mitani and J. Watarai found that the active material of the bacterium would dissolve in

Shigetane Ishiwata

alkaline solutions (but not in water alone) and pass through a filter, suggesting that it was an alkali-soluble toxin. Two other Japanese scientists, Kiyoshi Aoki and Y. Chigasaki, continued these studies and in 1911 observed that to kill silkworms, the culture must be old and must have produced spores. They confirmed Ishiwata's observations that larvae became paralyzed within sixty to eighty minutes after eating the spores. They were surprised to find that not all strains of the bacterium would kill silkworms. To make the story even more complicated, in 1926 another Japanese scientist found a *Bacillus sotto* strain that would kill insects other than caterpillars, including beetles, wasps, and crickets.

Far away in the Thuringia region of Germany in the summer of 1909, Ernst Berliner described a bacterium that killed a species of flour moth, *Ephestia kuehniella*, that was infesting a flour mill. Berliner named this bacterium *Bacillus thuringiensis*, in honor of the region where it was found. He observed that the disease caused the caterpillars to undergo a sudden paralysis or "sleep," and so he called the bacterium the "paralysis bacillus." Berliner noticed that the rod-shaped cells in the late stages of growth contained strange elliptical bodies that glittered brightly in the far end of the rod. He then proceeded to test whether the bacterium would kill mealworms (beetle larvae) as well as moths found in the granaries, but he did not find any effects on these beetles. Berliner was the first person to suggest that *Bacillus thuringiensis* might be useful in controlling pest insects.

Many important scientific findings are not appreciated at the time of their discovery, and unfortunately Berliner's original strain of *Bacillus thuringiensis* was lost. However, a decade later, another German scientist, Otto Mattes, was able to isolate the organism again from the flour moth and described similar symptoms in the insects. This strain was eventually passed on to scientists in the United States and Europe, where it led to the first attempts to use a bacterium for control of pest insects.

During the late nineteenth and early twentieth centuries, American and European scientists experimented with a number of insect pathogens, especially fungi, as potential biological control agents against a variety of insects. Success was at best erratic—some of these agents appeared to work and others failed. Gradually the agricultural community came to the conclusion that using microorganisms for control of insects was simply not practical. But *Bacillus thuringiensis* and the efforts of several dedicated scientists would soon change that attitude.

The first successful target for *Bacillus thuringiensis* in agriculture was the European corn borer, *Pyrausta nubialis*. In 1927 scientists from southeastern Europe and North America met to discuss their efforts to control this serious pest, and some reported that *Bacillus thuringiensis* and one fungus seemed to be potentially useful. At the same time, *Bacillus thuringiensis* was also being used successfully against the gypsy moth and several other caterpillar pests in France. Scientists at the Pasteur Institute in Paris took an active role in these studies. Serge Metalnikov and his son Serge Metalnikov Jr., who had recently escaped from the Bolshevik Revolution in Russia, isolated many bacteria from dead and dying caterpillars and explored the usefulness of *Bacillus thuringiensis* against corn borer, cotton bollworm, and several other species with encouraging results. The first commercial product based on the bacterium, Sporein, was produced and sold in France in 1938, for control of the bacterium's original host, the flour moth. More products entered the market in Eastern and Western Europe during the 1950s. But none of these products was used widely, and interest in insect control using microorganisms died away again for another decade.

One mystery that plagued attempts to use *Bacillus thuringiensis* against agricultural pests was its unpredictable activity. Many different strains were discovered in several species of caterpillars. Some of these were active against their original host, but some were not—a pattern that suggested that there were many different types of *Bacillus thuringiensis* and that each type killed a specific kind of insect.

As the number of new strains and interest in this bacterium grew, scientists needed to come to some agreement on what to call it. In 1951 Constantin Vago and Constantin Toumanoff isolated a bacterium from silkworms suffering from "infectious flacherie disease" at the Pasteur Institute. They named this bacterium *Bacillus cereus* var.

alesti because it was similar to the well-known microorganism *Bacillus cereus*. Toumanoff and Vago recognized that the symptoms and the bacterium were very similar to those described by Ishiwata and colleagues in Japan fifty years earlier. After a careful comparison of *Bacillus cereus* var. *alesti*, *Bacillus sotto*, and *Bacillus thuringiensis*, they concluded that all three "species" should be designated as *Bacillus cereus* and should thereafter be called varieties of the same species. Toumanoff and Vago declared that most if not all of the strains of bacteria with similar physical and chemical characteristics (whether they were isolated from caterpillars in Europe, Japan, or North America) were probably the same species. They also confirmed that these strains were different from all other insect pathogenic bacteria. After many publications, considerable discussion, and some heated arguments, *Bacillus thuringiensis* has been retained as the official name for these bacteria (although there is still general agreement that *Bacillus thuringiensis* is very similar to *Bacillus cereus*), and the abbreviation BT is now commonly used for the species.

In North America, interest was rising in the potential of microorganisms for control of agricultural and forest pests. In 1945, a young scientist, Edward Steinhaus, had just begun his career. As a graduate student and postdoctoral associate at Ohio State University, Steinhaus had molded his unusual combination of interests in bacteriology and entomology into the study of microorganisms associated with insects. He was hired at the University of California at Berkeley to establish the first laboratory of insect pathology in the United States. Steinhaus immediately began to train young scientists, offering the first course in insect pathology at Berkeley in 1947 and building a laboratory that would eventually have a major impact on the field. He traveled to many countries, including the Soviet Union during the cold war era, to develop international collaborations. He published a set of books on insect pathology in 1949 and 1963 that became critical to the development of the field. Steinhaus brought together a group of scientists to found the Society for Invertebrate Pathology in 1967, and was the first editor of the *Journal of Invertebrate Pathology*. As a result of his many efforts in the field, he is regarded as the father of modern insect pathology.

Steinhaus isolated bacteria from caterpillars and began to consider whether these bacteria might be useful agents for control of alfalfa caterpillars, a major pest in California. A colleague, Nathan Smith, identified these bacteria as strains of *Bacillus thuringiensis*. Because of his international connections with scientists in Europe and Japan, Steinhaus was well aware of the history of this bacterium and its rapid action. In one of those serendipitous events that sometimes happen in science, one strain he tested was the original strain that Mattes had isolated from the flour moth in Germany in 1911 (a strain that had been passed from scientist to scientist across Europe and the United States). The culture had been held in the refrigerator for a number of years. Steinhaus and his coworkers were amazed to find that the old Mattes culture killed the caterpillars overnight, while fresh cultures recently prepared in the lab did not. After assuring themselves that they had not contaminated this preparation

Left to right: Edward Steinhaus (1914–1969), John Briggs (1926–2002), and Constantin Toumanoff (1903–1967) at International Colloquium on Insect Pathology and Biological Control, Prague, 1958

with a chemical, they determined (as had Ishiwata) that insecticidal
activity was associated with older, sporulated cultures.

Steinhaus immediately began to investigate the usefulness of this bacterium for control of insects in the field and to talk industry into making BT products for sale to farmers. His efforts resulted in the first commercial BT product in the United States, Thuricide, produced by Bioferm Corporation in California. The U.S. Food and Drug Administration, after examining extensive tests of this product, declared it exempt from the strict rules applied to chemical insecticides, and in 1961 Thuricide was registered for use.

Commercialization of BT insecticides was not easy. The early products killed only caterpillars and not other pest insects. They were difficult to apply, they were difficult to preserve for long-term storage, and their activity was unpredictable. Moreover, the companies that produced chemical insecticides were generally not interested in these products. Why should they bother with microbials when chemicals were easier to produce, often were less expensive, lasted a long time after spraying in the field (especially organochlorides such as DDT), and killed nearly every insect they contacted? Moreover, there was concern that BT might kill beneficial insects as well as the targeted pests. This was particularly a worry in Japan, where the silk industry was an important source of income. The Japanese government banned importation of BT except for experimental use.

Then in 1962 Rachel Carson's *Silent Spring* awakened the world to the side effects of some chemical insecticides. Populations of hawks and other birds of prey crashed because of DDT-induced thinning of their eggshells, and it began to appear that human health might also be harmed by these insecticides. The "green revolution" was born. Also, many observers had noted that when the same chemical insecticide was used year after year on the same fields, increasingly higher doses were needed to control the insects. One by one, insect populations built up resistance to these chemicals, as the small percentage of survivors from the spray became the parents of the next year's insect population, carrying genes for resistance to the next generation. These events made microbial insecticides look much more attractive, with their narrow range of kill, no evidence of effects on living things other than the target insect, and rapid breakdown in the environment. They were considered to be environmentally friendly pest control agents.

Japanese scientists also recognized the potential for BT as an insecticide, but their government had prohibited its use. A group of scientists led by Keio Aizawa at Kyushu University decided that the solution was to find a strain with low activity against the silkworm but high activity against pest insects. Eventually, after considerable searching and selection, they found such a strain. Through extensive experiments, Aizawa and his colleagues found that careful use of this strain did not have a severe effect on sericulture, and in 1971 the government approved use of these BT products in Japan.

Bacillus thuringiensis products were tested against forest and agricultural pests during the 1960s, but sadly the results were not encouraging. Sometimes the product was a success, but at other times it failed, and the activity against different species of caterpillars was unpredictable. Then, in 1970, Howard Dulmage, a scientist with a U.S. Department of Agriculture laboratory in Texas, reported the discovery of a strain from pink bollworms that had a high level of activity against many different types of caterpillars. This strain, named HD-1, quickly became the most common strain in commercial products designed for control of caterpillars. Dulmage also developed a system for determining the insecticidal activity of commercial BT products, based upon a carefully preserved international standard. This standardized assay method became widely used, and product activity in international units now appears on the side of the package of any BT product.

So this brings us back to the question, how does BT kill caterpillars so quickly if (as Ishiwata observed in 1901) it does not multiply in the caterpillar until after the caterpillar dies? And why are different strains of BT so unpredictable in their activity? Many scientists—beginning with Ishiwata, Berliner, Mattes, and Steinhaus—had observed that BT exhibited its rapid activity only if the culture had already produced spores and that many of these spores lay at a slant inside the rod-shaped bacteria. Indeed, this feature was so

consistent in insect-pathogenic strains that it was one of the characteristics used to identify a new strain as BT.

In 1953, a Canadian scientist, Christopher L. Hannay, published
photographs of a diamond-shaped crystal, or "parasporal body,"
that lay alongside the spore in the BT cell. Hannay isolated the
crystals from the spores, dissolved them in alkaline sodium hydrox-
ide solution, and speculated that there was a relationship between
these inclusions and pathogenicity of the bacterium. A year later,
another Canadian, Thomas Angus (working on his Ph.D. research

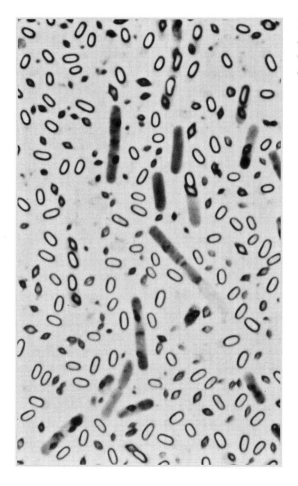

Light micrograph of
Bacillus thuringiensis,
demonstrating spores
and crystals

at the Canadian government's insect pathology laboratory in Sault Ste. Marie, Ontario), demonstrated that a toxic substance could be extracted from sporulated cultures using either gut extracts from caterpillars or an alkaline buffer. This toxic material caused rapid paralysis and death when fed to caterpillars. Angus observed that only the crystal, and not the spore, dissolved in alkali or gut extracts, so the active material appeared to be in the crystal.

But what were these strange crystals? To examine them better, Angus and Hannay turned to a new instrument, the electron microscope. No matter how powerful the lens, the ability of a light microscope to distinguish very small objects is limited by the wavelength of visible light. The electron microscope, developed in the 1930s, uses electron beams that have a much shorter wavelength than that of visible light and thus are capable of distinguishing between two objects that are very close together or very small. However, because these beams will bounce off air molecules, the electron microscope operates in a vacuum, and since our eyes cannot see electrons, we must visualize them on a screen much like a black-and-white television. Cells must be treated with special chemicals and sliced very thin so the electrons can pass through and form a picture on the screen. All of these techniques were being developed during the 1940s and 1950s, and the early electron microscopes were large, difficult instruments to use.

Angus, Hannay, and others discovered that when viewed under the electron microscope, crystals of various sizes and shapes were in all BT strains, again suggesting that these crystals were important to the insecticidal activity of the bacteria. Ishiwata's observation that activity seemed to be "in or on the spore" was indeed prophetic — it was in the nearby crystal.

A major question still remained, however: why are different strains of BT widely different in insecticidal activity? If there is only one crystal toxin, why is there a large variation in the susceptibility of insects to this toxin? Many scientists around the world studied different strains of BT during the 1980s, using increasingly sophisticated biochemical techniques. The crystal was found to be composed of a large protein that dissolved in the alkaline gut of caterpillars. This discovery led Arthur Heimpel, working with Tom Angus at the Canadian laboratory, to speculate that some caterpillars were not susceptible to the toxin because their gut was not sufficiently alkaline to dissolve the crystals. When the crystal entered the gut of susceptible caterpillars and dissolved, releasing the large protein, gut enzymes quickly clipped it into a smaller molecule, which was the real toxin. This molecule was named the delta-endotoxin. Microscopic studies revealed that once the delta-endotoxin entered the gut of a susceptible caterpillar, the gut cells swelled rapidly and soon burst. Leaked gut contents led to the rapid paralysis and death that Ishiwata and others had observed many years earlier. The long collaboration and strong friendship of Heimpel and Angus at the Sault Ste. Marie laboratory was to lead to many significant discoveries, as they gradually teased out how BT kills its host.

As the commercial possibilities of BT became obvious, many more strains were isolated, and the problem of classifying them became a major concern. A company needed to be able to say exactly which strain was in its BT product. Angus and Heimpel came up with a key to identifying strains based on biochemical and physical characteristics, but differences of opinion still frequently occurred among scientists working with this bacterium. Finally, two systems were developed to catalog the strains into manageable groups. Both of these techniques involved raising antibodies in rabbits that were specific to certain clusters of strains.

At the Pasteur Institute, Huguette de Barjac and her colleagues developed a way to classify strains according to the antibody reactions of the bacterial flagella (thin protein fibers on the outside of the cell that allow it to swim). De Barjac washed flagella from the surface of individual bacterial strains and injected them into rabbits. When sufficient time passed to permit the production of antibodies to the flagella, the rabbit's blood was taken and the liquid serum separated from the cells. This serum was placed in a well on an agar plate, with flagella from the BT strains placed in surrounding wells. When these proteins migrated through the agar and met each other, a characteristic set of visible bands appeared. De Barjac discovered that the BT strains could be grouped using this technique, which she called serotyping.

Serotyping added significant order and reproducibility to the cataloging of the strains. However, researchers quickly discovered that the flagellar serotype did not always predict insecticidal activity. To sort out this problem, another antibody-based identification system was developed for the crystal proteins by Jenina Krywienczyk at the Sault Ste. Marie laboratory. Finally, Howard Dulmage directed an international collaborative program that clearly demonstrated that the insecticidal activity of BT strains was related to the crystal serotype. Apparently, the crystal toxin proteins were sufficiently different that unique antibodies were produced for each one. Both serotyping systems are still routinely used to classify new strains. There are currently more than sixty different types of BT, containing more than a thousand strains.

Not all important agricultural pests are caterpillars, and many other insects, especially beetles, were also candidates for biological control. Until the 1980s, beetles were thought to be immune to the activity of BT, which appeared to be limited to butterflies, moths, flies, mosquitoes, and their relatives. However, in 1983, two BT strains were separately discovered in Germany and California that were highly insecticidal for beetle larvae. Three German scientists —Aloysius Krieg, Alois Huger, and Wolfgang Schnetter—eventually showed that the two strains are identical. This BT strain is very effective against certain beetles, such as the Colorado potato beetle and, in particular, the cotton boll weevil. As new strains of BT have been discovered, a remarkable range of pests have been found to be susceptible, including cockroaches, mites, lice, stinkbugs, biting midges, scarab beetles, and even nematode worms and fungi. The full potential of BT as an insecticide has not yet been realized.

Molecular biology techniques developed rapidly in the early 1980s, enabling scientists to answer the next important question, what genes lead to the production of the delta-endotoxin? In 1982 Jose Gonzalez and Bruce Carleton discovered that the genes coding for the toxins of BT reside on small circular bodies of DNA called plasmids. Bacteria frequently exchange plasmids with each other, and these small loops of DNA are known to confer many important properties to bacteria, including resistance to some antibiotics. When the toxin genes were found on plasmids, their presence accounted for some of the complexity of toxins found in BT and explained why strains with one type of flagellum could have several different types of toxins. The findings also could be applied to explain how some of the most useful strains (such as Dulmage's commercially valuable HD-1) were active against many different species of caterpillars while other strains were much more restricted in their host range. HD-1 was eventually found to carry a number of toxins; nature had genetically engineered this organism by adding several plasmids, each coding for a different toxin.

H. Ernest Schnepf and Helen Whitely at the University of Washington soon succeeded in cloning the first toxin gene from HD-1 into the common gut bacterium, *Escherichia coli*. Many other delta-endotoxin genes were soon cloned, and in 1985 several laboratories reported the first sequences of these genes. When many gene sequences were compared, patterns began to appear. Currently delta-endotoxin genes are divided into five major groups based upon their genetic content and their activity against insects.

Once the toxin genes were cloned, the race was on to understand how the toxin actually kills the insect. To understand with this process, we must know several things: the shape of the toxin in its active form; how it first attaches to the cells of the insect gut; and finally, how it interacts with these cells to kill the insect. Several laboratories set out to tackle these questions.

In 1991, Jade Li, Joe Carrol, and David Ellar at Cambridge University in England described the first three-dimensional structure of a BT toxin, the delta-endotoxin active against beetles. Solving the three-dimensional structure of a large protein molecule is a major undertaking, involving purification of the protein, inducing it to form a crystal, exposing the crystal to X-ray beams, and analyzing the pattern of rays bouncing off the layers of amino acids within the crystals. This tedious process often takes years. Combined with analysis of genetic mutations of the toxin genes, these studies eventually revealed that part of the toxin protein was responsible for toxicity and a different part was responsible for host specificity.

Meanwhile, studies at the Cambridge laboratory and elsewhere were beginning to explain the rapid activity of the toxin in the insect. After the crystal dissolves in the gut of the insect and is cleaved by enzymes to its active form, it attaches to specific sites on tiny finger-like structures that extend from the gut cells into the center of the gut. These sites, called receptors, were found to be proteins with attached sugars that vary from one insect to another, thus explaining the specificity of the toxins for certain insects. There must be an exact match between the receptor on the gut cells and the portion of the toxin responsible for specificity. Finally, researchers determined that the toxin molecules form a pore in the membrane, permitting gut fluid to leak into the cell, which forces the gut cells to rapidly expand and burst. These discoveries, all made during the decade beginning in 1985, answered many of the questions raised by scientists about the remarkable activities of BT over the previous 150 years.

BT products have been very successful in control of the notorious imported pest of forests, the gypsy moth. Thousands of acres of forest in the United States and Canada continue to be sprayed each year from airplanes with BT products. Field crops, however, present a different challenge. Although commercial BT products work very well and are not toxic or infectious to other animals, they have some limitations that make it difficult for them to compete with chemical insecticides. With the exception of the HD-1 strain, BT strains generally kill only a narrow range of target insects. Since the product has to be eaten by the insect to be effective, placing it where secretive species (such as the bollworm, which bores inside the cotton boll) can encounter it is difficult. Also, the activity of BT is rapidly inactivated by sunlight, whereas chemicals lose their activity more slowly. Because of these limitations, a farmer can use BT to control only one or a few kinds of insects on his crop, and he must spray it frequently. As a consequence, traditional sprayed BT products are used in only a small percentage of insect control efforts in field-crop agriculture.

Then in 1987 the first BT genes were inserted into tobacco and tomato plants, with the goal of having the delta-endotoxin reach the target insect if it eats any part of the plant. The toxin is present and

active for the life of the plant and reduces the use of expensive spray equipment. Transforming plants with BT genes has proven highly effective. In addition to tobacco and tomato, other crop species such as cotton, soybeans, corn, peanut, eggplant, sweet potato, rice, coffee, and even trees have been genetically engineered to contain BT toxins. In 1995 the first BT toxin–containing plant, field corn, was approved for sale in the United States. Currently, millions of acres of cotton, soybeans, corn, potatoes, and rice containing BT genes are planted in the United States. Many years of BT use in field crops have clearly demonstrated that it does not harm fish or other invertebrates aside from insects, and it is not toxic to beneficial insects such as bees. BT-engineered crops considerably reduce the use of insecticidal chemicals, and there is no evidence of harm to mammals or birds.

Although genetically engineered crops expressing BT toxins have been well accepted by U.S. agriculture, they have their own set of problems. Products aimed at killing caterpillars may, in some cases, kill caterpillars of butterflies that were not the target of the toxin. A report published in 2000 claimed that pollen from corn expressing BT toxin was lethal to monarch butterfly caterpillars in the laboratory. This report stirred great controversy in the scientific community and in the press, but later studies of the distribution of pollen and monarchs in the field demonstrated that commercial cultivation of BT corn did not pose a risk to monarch populations. Although there is no evidence of harm from these crops or from BT itself, concerns have been repeatedly raised that they may induce allergies in human consumers, have effects on pollinators and biological control organisms such as parasitic wasps, reduce biodiversity by replacing native crops, or transfer genes to other species. Great controversy has also arisen over the introduction of BT-modified food into the diet. In 2000, flour from BT-modified corn was found to have made its way into a brand of taco shells, leading to the recall of this product. Although this corn was approved for feeding to animals, it was not yet approved for human consumption, owing to the possibility of development of allergies.

A report produced in 2004 by the Commission on Environmental Cooperation (representing Canada, Mexico, and the United States) called for a moratorium on the planting of genetically modified corn in Mexico to avoid potential genetic crossing with native varieties, a suggestion that was hotly contested by the U.S. government. In Europe, the controversy has risen to a level where genetically modified, or GM, foods have been declared "Frankenfoods" by activists. Experimental fields have been destroyed by these groups, and in some countries the sale of GM foods has been totally banned. In 2004 and 2005, China pondered whether farmers in that country should grow BT-modified rice. This rice can reduce pesticide use by as much as 80 percent, but it is resisted by farmers who use traditional varieties. In perhaps the saddest case, BT-modified corn donated to Zambia in southern Africa to relieve widespread starvation in 2002 sat in warehouses because the country's leader would not permit GM corn to be given to the people. However, in 2005 BT-modified corn was approved for trial plantings in Kenya to stop a stem borer that causes the loss of thousands of tons of this critical crop. It remains to be seen how BT-modified crops, as well as other GM crops coming onto the market, will be accepted by the general public and governments in the future.

A realistic concern about BT modification of plants is the fear that widespread planting of these crops will lead to development of resistance to BT, as repeatedly occurs with chemical insecticides. Ironically, the major reasons for incorporating the toxin genes into plants were to get around rapid degradation of activity by sunlight and accumulation only on the top of the leaves by sprayed BT products. This attractive property of GM plants, long-lasting activity, may also promote the development of resistance. For an insect population to become resistant to BT (that is, requiring more and more product to kill each new generation), the product must be present throughout the season. When BT was a sprayed product based on the bacterium itself, resistance was not a problem. Indeed, the only report of resistance was in moths treated with BT in grain storage silos, where sunlight never reached the product. When William McGaughey reported this finding in 1985, it was met with surprise, as it was the first evidence of resistance to BT-based products in more than twenty years of use.

However, many more species were eventually shown to be capable of becoming resistant in laboratory experiments. Strategies were rapidly developed by seed companies and farmers to avoid development of resistance. These include the planting of non-BT crops along with those expressing the toxin to form refuges for insects that are not exposed to the toxin. This technique is based on the logic that these susceptible insects will mate with the survivors of BT-engineered plants, diluting any genes for resistance. Whether farmers and companies will accept this technique remains to be seen, however, as it may result in partial loss of crop from the refuges. New genetically modified plants are being created that contain more than one BT toxin strain and even toxins from bacteria symbiotic in insect-pathogenic nematodes. This technique, called pyramiding, is expected to slow development of resistance. By 2004, BT corn and cotton were planted in over 80 million hectares (approximately 200 million acres) worldwide, with no resistance yet documented. If resistance to BT toxins does develop, it will result in the loss of one of our best and safest insect control agents.

3

Out of Africa

Bacteria against Mosquitoes and Blackflies

Bacillus thuringiensis and its relatives have even more qualities that have the potential to improve the health of humankind. The following stories of onchocerciasis, malaria, and West Nile virus illustrate why these qualities are important.

> In his village in Burkina Faso, West Africa, Olatu was preparing his small field for planting. He had difficulty seeing his hoe and asked his small son to help him with the work. Olatu remembered the warnings of the elders when he moved his family to the un-inhabited valley near a beautiful fast-flowing stream. The elders had warned Olatu that the stream was evil and said that ferocious flies born in that stream were demons. They told him that people who lived there became blind. But his family had outgrown their small field in the uplands, and they desperately needed more land. Within a year, Olatu became blind. All of his children had lumps beneath their skin and spent much of their time scratching annoy-ing rashes. Their beautiful dark skin developed white spots; Olatu knew that the curse of the stream was upon them.

The curse of the stream is river blindness, or onchocerciasis, caused by a parasitic worm, *Onchocerca volvulus*, carried by several species of the blackfly genus *Simulium*. Blackflies breed only in fast-running waters, where their cigar-shaped larvae attach to rocks and spread fanlike mouthparts to filter bacteria and algae from the water flowing past. In West Africa, a new generation of blackflies emerges every eight to ten days year-round. They are ferocious biters, persistent and very annoying, leaving behind a large welt that often lasts for many days and can cause serious medical problems in allergic individuals. Blackflies are a problem in nearly all regions of the world where there is fast-flowing water, from Africa to Canada. Although North American blackfly populations do not carry human disease, they have a strong economic impact on tourism and outdoor activities when they are out in force in the spring and early summer.

Onchocerca adult worms lodge in lumps under the skin of infected people. There the female and male worms mate, and the female produces millions of microscopic larvae. These tiny worms migrate throughout the body in the blood and cause the symptoms that Olatu and his family experienced. When a person carrying onchocerciasis is bitten by a blackfly, the worms are carried by the fly to the next victim. The infected person can continue to spread the disease for up to fourteen years. As happened to Olatu, the worms often lodge in the eyes of the victim and lead to blindness. In the savanna areas of West Africa, river blindness has prevented entire fertile valleys from being used for farming; this disease is a major obstacle to economic development of the region.

On the other side of Africa, Anna Chickwamba smiled down at her baby sleeping in her lap, barely visible in the dark, smoky interior of her mud hut. Unlike her first daughter, this baby had survived for six months but was beginning to look pale and feverish this evening. Anna feared for her baby, as she had seen too many other children die of the terrible fever and convulsions that haunted their village on the edge of the beautiful Great Rift Valley in East Africa. This village, like many others in Africa south of the Sahara Desert, suffered from constant attacks of malaria.

Every thirty seconds, a person dies of malaria somewhere in the world. In Africa, South America, India, and Southeast Asia, malaria kills more than a million people each year, most of them children under the age of five. Even the United States is not free from the devastating effects of this disease. Before the introduction of modern insecticides, malaria (then called "fever and ague") was common across the United States. In the summer of 2004, seven people contracted the disease in Florida, the first time malaria has been transmitted within the United States in more than twenty years.

Malaria has been associated with humankind for as long as humans have existed. The disease probably originated in Africa, and it traveled right along with humans as we invaded the rest of the world. The early Romans named it malaria ("bad air" in Latin), but as early as 95 BC, Lucretius suggested that perhaps it was caused

Onchocerciasis victim: a sixty-three-year-old male farmer, virtually blind, his eye indicating the damage that occurs as a result of long-term infection

by a living thing, not just the air. Serious epidemics of this disease in Rome and across southern Europe were recorded in the first century AD. In the early 1700s, Giovanni Maria Lancisi, physician to the pope, noticed that when swamps near Rome were drained, malaria died away. Lancisi was the first to suggest that an insect from the swamps was somehow spreading this disease.

Malaria is caused by four protozoan parasites: *Plasmodium falciparum*, *Plasmodium vivax*, *Plasmodium ovale*, and *Plasmodium malariae*. Although *Plasmodium malariae* may infect other primates, all the rest are strictly diseases of humans. These protozoa were first identified by a French surgeon, Charles Laveran, in 1880. The other actor in the story is the mosquito, genus *Anopheles*. The connection between *Anopheles* mosquitoes and malaria was confirmed by Ronald Ross, who first saw the *Plasmodium* parasite in the gut of an Indian mosquito in 1897.

The *Anopheles* mosquito prefers to bite humans. If she (only the female mosquito bites) bites someone who is infected and whose

Anopheles mosquito biting, exuding fluids

parasites have reached the infectious stage, she will take up the malaria parasites along with her meal of blood. The parasites then infect the mosquito and move to her salivary glands; when she bites the next person, the parasite moves to a new host. In less than an hour, *Plasmodium* cells move out of the host's blood and into the liver. There they multiply and enter red blood cells. When the blood cells burst, expelling a new generation of parasites, the victim experiences the chills and fever that are characteristic of malaria. The parasite is waiting in the blood for the next mosquito bite. If not treated, a human can remain infectious to others for many years.

> Mark is a teacher in Phoenix, Arizona, in the northern reaches of the Sonoran Desert. When he came down with a terrible backache, splitting headache, stiff neck, and rash, he thought he probably had the flu. His doctor was not so sure, and he was tested for every disease that the doctor could think of—meningitis, valley fever, tuberculosis, pneumonia. None of the tests turned out to be positive. Mark's next-door neighbor, also a doctor, suggested that he should be tested for a new disease, West Nile virus. Five days later, Mark learned that this was the cause of his illness. Then Mark remembered seeing several dead crows at the golf course when he was enjoying his Saturday game and being bitten by a couple of mosquitoes near the golf course pond. After several weeks, Mark was able to return to teaching but was still very tired at the end of the day.

West Nile disease was first reported in 1937 in the West Nile district of the African nation of Uganda. When it invaded Egypt in the 1950s, it was discovered to be carried by mosquitoes, and a virus was confirmed to be the cause of severe meningitis and encephalitis in elderly patients during an outbreak of West Nile disease in Israel in 1957. Soon horses and birds were found to be infected as well. The West Nile virus made its way to the eastern United States in 1999. Since then, the disease has moved steadily westward with birds and horses. Several hundred deaths a year from West Nile virus are reported in the United States, generally in the very young and the elderly. But most persons who catch the disease do not have any symptoms and may become immune.

The culprits in West Nile disease are birds that harbor the virus and move it during migrations, coupled with mosquitoes of the genus *Culex* that transmit it among birds and to humans and horses. *Culex* is a very common mosquito that is closely associated with humankind almost everywhere in the world. *Culex* mosquitoes take advantage of convenient sources of water provided by humans to breed. The larvae, or "wigglers," can be found in any small puddle that stays full for two weeks or so—a dog's water bowl, a child's wading pool, a flowerpot, a swamp, or a neglected swimming pool. Like blackfly larvae, mosquito larvae feed by spinning fanlike mouthparts to collect bacteria, algae, and other bits of nutrient material from the water. They hang from the surface of the water by a "snorkel" tube that allows them to breathe while feeding beneath the water surface. When they complete their larval cycle, they form round pupae. Within a couple of days, the pupa floats to the surface and an adult mosquito pops out and flies away to seek a blood meal. If a female *Culex* takes a blood meal from an infected bird, the virus undergoes a multiplication cycle in the mosquito and makes its way to her salivary glands. A week or so later, her next blood meal can result in transmission of the virus to a bird, horse, or human.

Eradication of these devastating diseases has been a goal of national and international agencies for many years. Effective chemical insecticides, beginning with DDT, began to appear in the 1940s. A very successful effort to eradicate malaria was initiated in the United States in 1947 using chemical insecticides. In an optimistic attempt to duplicate the American results, the World Health Organization (WHO) launched the Global Malaria Eradication Program in 1956. However, the disease and its mosquito carriers were not easily conquered in the rest of the world. The *Plasmodium* parasites rapidly developed resistance to antimalaria drugs, and widespread use of DDT led to the development of resistant *Anopheles* mosquitoes in several regions of the world. These discouraging results led to abandonment of WHO malaria eradication attempts.

Beginning in 1962, however, the WHO decided to try a different approach and to look for biological control organisms that were specific for insects, called vectors, that carry diseases. The Tropical Disease Research Working Group was initiated to provide funding, which encouraged scientists to investigate this new idea. A WHO Collaborating Center for Biological Control was established at Ohio State University, under the direction of John Briggs, a former student of Edward Steinhaus. This center collected potentially useful organisms sent from scientists around the world and distributed them for experiments to determine their effectiveness. Malaria and onchocerciasis were major targets of these efforts, and any organism that would kill a mosquito or a blackfly was placed under intensive study.

The search led to the discovery of several fungi, viruses, protozoa, and nematode worms that were potentially useful, but these organisms proved difficult to produce in large quantities and to use in the remote, tropical areas where they were most needed. Many strains of *Bacillus thuringiensis* were tested but had only a low level of activity against mosquito larvae. A related bacterium, *Bacillus sphaericus*, found by William Kellen in sick mosquito larvae in California, also proved to be only mildly effective. The prospects for using microorganisms to control disease-causing insects did not look promising.

Then, in 1970, a collection of dead mosquito larvae was sent from the Ohio State collection to Sam Singer at Western Illinois University. From these larvae, Singer isolated a new strain of *Bacillus sphaericus* that he called SSII-1. This strain was significantly better at killing larvae than was Kellen's strain. Although SSII-1 was very unpredictable in its activity, it did point to the possibility that more useful strains of *Bacillus sphaericus* might be waiting to be discovered. In early 1975, Singer isolated another strain, designated 1593, from WHO material that confirmed that *Bacillus sphaericus* had great potential for control of mosquito larvae. Strain 1593 was stable over long-term storage and was highly active against all *Culex* larvae. Singer and his collaborators found that the bacterium grew well on common laboratory media, and one milliliter, only about twenty drops of culture, contained enough bacteria to kill almost a million *Culex* larvae. Strain 1593 was lethal to some *Anopheles* species as well but was disappointingly weak against *Aedes aegypti*, the other major target, which carries yellow fever and dengue fever. Eventu-

ally, more than thirty insecticidal strains of *Bacillus sphaericus* were
under study in laboratory and field trials around the world, and a
Nigerian strain called 2362, isolated by the Czechoslovakian scientist Jaroslav Weiser, proved especially promising.

Bacillus sphaericus was found to kill mosquito larvae with a toxin
composed of two proteins, both of which are required for activity.
Like BT, *Bacillus sphaericus* produces this toxin when the rod-shaped
vegetative bacterium shifts to forming spores. Also like BT, the *Bacillus sphaericus* toxin is concentrated in a tiny crystal lying alongside
the round spore. When the spore and toxin crystal are eaten by a
mosquito larva, the crystal dissolves within a few minutes and be-

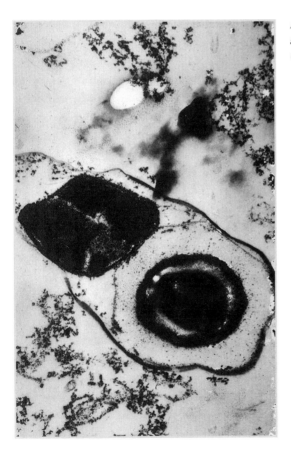

Bacillus sphaericus spore
and toxin crystal
(electron micrograph)

gins to destroy the gut cells. The larva stops feeding and dies within a day or two, and the bacteria multiply in the corpse. Each mosquito larva killed by *Bacillus sphaericus* can yield thousands of new bacteria within a few days, ready to infect the next generation of larvae in the pool.

Bacillus sphaericus remains active in the environment for up to a month after application, a very beneficial trait in the context of small pools that frequently dry up and refill again at the next rain. Unlike that of BT, the toxin of *Bacillus sphaericus* appears to be protected from the degrading activity of the sun by the remains of the bacterial cell wall, accounting for its persistence in the larval breeding pools. The ability of this bacterium to wait patiently in the mosquito habitat for the next generation of larvae was an important benefit. Extensive research proved that *Bacillus sphaericus* is safe for use around mammals, birds, and fish and does not kill beneficial organisms. These results led to the development of commercial *Bacillus sphaericus* products that are currently in use for mosquito control in the United States, Brazil, and several other countries where diseases vectored by *Culex* and *Anopheles* are serious threats to human health. Such products are also used to control mosquitoes that vector the West Nile virus in several states in the United States.

But yet another bacterium was to have an even greater impact on insect vectors of human disease, and it was found in a very unlikely place. The Negev Desert that makes up about 60 percent of the state of Israel is usually dry and desolate, full of sand dunes and mountains, inhabited mainly by Bedouin tribespeople and their sheep and camels. Nonetheless, after the winter rainy season, small mosquito-breeding pools remain in the single stream running through the region. Joel Margalit, a professor at Ben Gurion University of the Negev in Beer-Sheva, surveyed many small ponds in the Negev Desert looking for sick mosquito larvae. In the summer of 1976, Margalit observed large numbers of dead and dying *Culex* larvae in a small salty pool full of sheep and camel manure. He took a sample of the dead larvae, along with water and mud, and returned to his laboratory at the university. There Leonard Goldberg, a visiting scientist from the U.S. Office of Naval Research, isolated hundreds of spore-forming bacteria from the samples, but only one had any activity

toward mosquito larvae. Goldberg called that strain ONR-60A.

   In the next few months, Goldberg and Margalit tested strain ONR-60a against *Culex*, *Aedes*, and *Anopheles* larvae. It turned out to be amazingly active against every species they tried. Cultures were sent off to the Ohio State collection, to the WHO in Geneva, and to the Pasteur Institute in Paris. As news of this discovery spread

Joel Margalit collecting soil samples in the Negev Desert at the site of discovery of *Bacillus thuringiensis israelensis*

among the small circle of insect pathologists, an explosion of activity followed.

Huguette de Barjac at the Pasteur Institute received one of those cultures. De Barjac was the expert on *Bacillus thuringiensis* who had developed the technique called serotyping for identifying and classifying these strains. She quickly identified the Israel strain as a new serotype (H-14) and named it *Bacillus thuringiensis* (variety or subspecies) *israelensis*. De Barjac and her colleagues at the Pasteur Institute launched an intensive program of testing this new strain against mosquitoes and other insects and soon confirmed Goldberg and Margalit's results. *Bacillus thuringiensis israelensis*, or BTI as it quickly became known, was found to kill many species of mosquito larvae that vectored diseases. And what is perhaps even more encouraging, it also rapidly killed blackfly larvae. De Barjac described the discovery of the new Israel strain as "a little like the fairy tale, Sleeping Beauty." Interest in *Bacillus thuringiensis* as a bacterial insecticide had become dormant by 1977, but the discovery of this extremely active new strain led to a reawakening of interest in the possibilities of all BT strains for biological control.

As scientists began to explore why BTI could kill many different species of mosquito larvae, blackfly larvae, and even a few related insects, a complicated story began to emerge. As is true for all *Bacillus thuringiensis* strains, BTI produces a crystal alongside the oval resting spore. But when scientists began to look at this crystal under the electron microscope, they found there was not just one uniform crystal, but rather at least three apparently different parts. When chemical analysis was performed on these crystals and toxin genes were eventually cloned, researchers found that BTI produces at least four different toxins, plus two other proteins that add to the activity of the insecticidal proteins when they are all eaten together by a larva. This complex arsenal of toxins explains both why BTI can kill a large number of mosquito and blackfly species, and also the rapid lethality of the bacterium. When a larva eats enough of these crystals, it will stop feeding almost immediately and die within a few hours. Examination of these larvae with the electron microscope explained this rapid action; the gut cells looked like they had exploded. But unlike *Bacillus sphaericus*, BTI does not seem to multi-

The two proteins that enhance the activity of the insecticidal toxins were found to have another disturbing property, however: they cause mammalian red blood cells to burst. Extensive tests against mammals, birds, and fish eventually demonstrated that BTI is not hazardous at the concentrations used in the field. And it is much safer for the environment than chemicals since BTI does not kill bees, aquatic insects other than mosquitoes and blackflies, or other nontarget insects. Companies raced to develop products based on BTI, including liquid formulations, dry powders, and floating donut or briquette-shaped products. These briquettes can be dropped into a fishpond or flowerpot, where they float and slowly give off BTI to control mosquitoes breeding in your backyard. BTI is also a valuable component of West Nile virus control in the United States, where it is used in wetlands and lakes where the *Culex* mosquito breeds.

In 1974, in an attempt to reduce river blindness, the WHO launched an onchocerciasis control program in West Africa. Because infected humans can be reservoirs for this worm for up to fourteen years, the WHO realized that the blackfly must be controlled for more years than the humans retain the infection. This very ambitious project involved treatment of rivers and streams in eleven countries, over a region of 1.3 million square kilometers (800,000 square miles), inhabited by 30 million people. In this tropical region, there are nearly 50,000 kilometers (30,000 miles) of flowing rivers and streams. The WHO was joined by the World Bank, the United Nations Development Project, the Food and Agriculture Organization, and a coalition of twenty donor countries to sponsor this project.

As soon as BTI was shown to kill blackfly larvae, it was rapidly explored for use in the onchocerciasis program, but effective application was complicated by a number of obstacles. During the tropical rainy season when the rivers are running rapidly, BTI poured into a stream at one point raced past the blackfly larvae before their spinning mouthparts could filter out enough of the BTI to kill them.

In the dry season, the rivers become a series of ponds, and each pond had to be treated individually. Early formulations of BTI were active for only a few days in the field, so BTI had to be constantly added to these dry-season ponds. And there was the concern that the blackflies, which reproduce rapidly, would develop resistance to BTI if exposed over and over again. These were all challenges faced by the onchocerciasis control program.

The first of these challenges, formulation of BTI so that it would stay around long enough to be eaten by blackfly larvae, was approached by several companies. After trials of many different products, small particles of bacteria blended in liquid form turned out to work well in the fast-flowing rivers. Another challenge was stability of the product, since tropical heat could quickly kill it. New formulations were made to survive storage near the targeted rivers, where the BTI products were stored before being applied by airplanes. And formulations that killed mosquitoes did not necessarily work well against blackflies. After a great deal of effort by governmental agencies and commercial companies, a useful BTI formulation for the West Africa project was finally developed. BTI and chemical insecticides were alternated during the wet and dry seasons to get around the potential resistance problem.

But the final nail in the coffin of the river blindness worm was ivermectin, a new drug developed in the 1980s. Ivermectin proved safe and effective at killing the nematode worm within the human body. This drug has been supplied free by the manufacturer, Merck, which has donated millions of doses for annual treatment of persons living in onchocerciasis regions. The number of people infected with this disease in West Africa has decreased dramatically. In 1992, another onchocerciasis elimination program was begun in South America, and a similar program was established in 1995 for African countries not included in the original program. The initial WHO West Africa Onchocerciasis Control Program was declared a success and brought to an end in 2002.

BTI and *Bacillus sphaericus* continue to be important components of local and national vector control programs in the United States and Europe. An excellent example of the benefits of these bacteria is the project to control mosquitoes in the upper Rhine Valley in Ger-

many. In this carefully organized project, mosquito breeding sites
are all mapped, breeding habitats for fish, bats, and other predators
of mosquitoes are provided, and BTI is applied by trained workers.
The populations of mosquitoes and other organisms in the sites are
carefully monitored throughout the season. This project has been a
great success, with the result that local citizens are no longer assailed
by huge numbers of mosquitoes. The technology developed by the
German program was transferred to a region in China, resulting in
a significant reduction in malaria within three years.

Malaria, however, remains a challenge to humankind in many
parts of the world. Both BTI and *Bacillus sphaericus* are being tested
or are already in use against *Anopheles* larvae in countries where
this disease is still a major killer. Other important developments,
such as insecticide-treated bed nets and treatment of the inside of
houses with insecticides, have produced good results in controlling
the adult mosquitoes. New drugs, including one from a Chinese
herb, have potential for curing infected individuals and prevent-
ing infection. The full genetic sequences for the *Anopheles* mosquito
and the *Plasmodium* malaria parasite are now known. Understanding
these genes and how they interact may lead to new ways to control
this devastating disease.

As we have learned many times, when any insecticide is used repeatedly, the insects may become resistant. Resistance to BTI has
not yet arisen as a major problem, but *Culex* mosquitoes have become resistant to *Bacillus sphaericus* after continuous exposure for as
little as four generations. Scientists have approached this challenge
by combining the toxins from BTI and *Bacillus sphaericus* into genetically modified bacteria that have characteristics of both organisms
to overcome resistance, at least temporarily.

Two other challenges stand in the way of widespread use of these
bacteria for control of disease-carrying insects. Although *Bacillus
sphaericus* remains insecticidal for many weeks in the field, it tends
to settle within a day or so to the bottom of the pond. This places
it out of reach of the hungry *Culex* and *Anopheles* larvae feeding on
the surface. In contrast to *Bacillus sphaericus*, BTI loses its activity

rapidly after it is applied. An ambitious goal is to add the broad host range of BTI to the long-term persistence of *Bacillus sphaericus*. Recombinant bacteria have been produced by moving genes from one bacterium into the other, in an attempt to combine the best features of both. Genes for the BTI and *Bacillus sphaericus* toxins have also been introduced into algae and other floating organisms that are commonly eaten by mosquito larvae, in an attempt to overcome the problem of bacteria settling out of the feeding zone of the larvae. These and many other projects by ambitious scientists should extend the usefulness of these two remarkable bacteria, ultimately leading to fewer human cases of malaria, onchocerciasis, West Nile virus, and other insect-borne diseases.

4

The Virus That Cures

The Baculovirus Gene Expression System

We return to the silkworm for another story, much of which comes from the careful analysis of papers and texts — some almost five hundred years old — in several languages, compiled by the Swiss scientist Georg Benz.

Benz tells us that after the silkworm was brought to Europe around AD 555, farmers in many countries recognized a disease that they called by several different names: jaundice (in England), *grasserie* (in France), *giallume* (in Italy), and *gelbsucht* or *fettsucht* (in Germany). In 1527 the Italian poet Marco Girolamo Vida described a similar disease, causing the death and melting of silkworms. Vida described swelling or wilting of the insect's body, a foul smell, and putrid gore that flowed from the body. Maria Sibylla Marian later recorded the same set of symptoms, which she called *gelbsucht*, in her book on caterpillars written in 1679.

But it wasn't until 1856 that two Italian scientists, E. Cornalia and A. Maestri, suggested that tiny bodies that they could see in the cell nuclei of silkworms were associated with the jaundice disease. These minute bodies were multisided or polyhedral in shape and looked like tiny crystals in the nucleus of blood cells and other tissues. Significantly, the polyhedral crystals were always present when the silkworms had the disease. These crystals became the object of considerable debate. Some scientists thought they were part of a protozoan, others thought they came from a bacterium, and still others favored different microorganisms. During this debate, J. Bolle reported that these crystals dissolved in the juices of the gut when they were swallowed by a silkworm. Bolle observed a central granular mass and a thin outer membrane in the dissolved crystals. In 1906 he used a well-designed experiment to clearly prove that the polyhedral bodies were indeed the infectious agent of jaundice. He found that when silkworms were fed or inoculated with polyhedra, they always became diseased, but if the polyhedra were killed by heat, the caterpillars remained healthy.

Although other scientists agreed with Bolle that the disease seemed to be associated with the polyhedral bodies, there was still no agreement about what kind of pathogen this could be. Several scientists attempted to answer this question by using a simple filter that permits viruses (which are generally very small) to pass through, while larger bacteria, fungi, and protozoa are held back. Several early experiments appeared to show that the infectious material was retained on the filter and so could not be a virus. Finally, however, in 1913 Rudolf Glaser and J. W. Chapman at the Rockefeller Institute in the United States were able to show that the pathogen causing a disease of the gypsy moth—a pathogen that also produced polyhedral bodies—did, indeed, pass through a filter. The pathogen was therefore a filterable virus. The earlier confusion probably arose from concentrated suspensions of insect bodies clogging the filters. At that time, however, the term *virus* was used loosely to refer to almost any infectious agent; not until the 1920s was the disease widely accepted to be caused by an agent that today we would all agree is truly a virus.

In 1924 a French scientist, André Paillot, used wonderful new dark-field and phase-contrast microscopes to see much more detail in cells of infected insects. Paillot observed virus multiplication in the nucleus and discovered strange particles that appeared to form a ring within the nucleus of cells infected with the virus. In other caterpillars, Paillot discovered a different type of virus in the nucleus. These viruses produced much smaller, granular particles, barely visible even with his fine microscopes. Later Edward Steinhaus would name these "granulosis viruses." Paillot produced the first book to be devoted completely to diseases of insects, *L'infection chez les insects* (*Infections in Insects*), in 1933. This important book inspired many scientists to explore this new field.

One of these scientists, Gernot Bergold, studied diseases that he called polyhedral viruses in caterpillars in Germany from 1942 to 1948. Bergold and his associates described some of the biochemical properties of proteins from the virus particles and published the first electron micrographs of virus particles in 1943. Bergold found that there were rod-shaped bodies inside the crystalline polyhedra. This came as a surprise to scientists studying these diseases, who thought that the crystals themselves were the viruses. Bergold correctly concluded that these rods were not bacteria or some other microorgan-

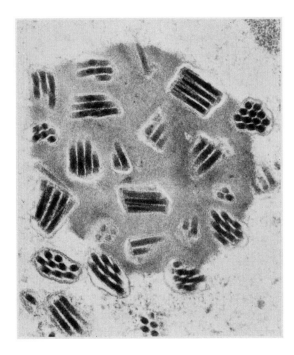

Nuclear polyhedrosis
virus (electron
micrograph)

ism but viruses hidden within the polyhedra. The viruses induced the insect cells to form the polyhedra, which provided a protective coat so that the infectious rods could resist sunlight, cold, and drying. The polyhedral body permitted the virus to remain in the environment from one season to another, waiting to infect the next generation of caterpillars.

As many more insects infected with viruses were examined by electron microscopy, three different types of polyhedral insect viruses were identified. The viruses producing large polyhedra in the nucleus of the cell were termed nuclear polyhedrosis viruses, or NPVs. The smaller, granular viruses, first described by Paillot, were called granulosis viruses, or GVs. Electron microscopy revealed that NPVs and GVs are similar; both produce containers, or capsids, made of one or more proteins, with viruses embedded inside. NPVs contain one to many rod-shaped virus particles within each large polyhedral capsid, while GVs contain a single rod-shaped virus within a small oval capsid. And finally, naked, nonoccluded rod-shaped viruses without capsids can also occur. All these viruses were found to contain circular double-stranded DNA as their genetic material and were named *baculoviruses*, from the Latin word for "rod" or "walking stick." Many different types of baculoviruses have been discovered, infecting more than six hundred species of insects, primarily moths and butterflies in the caterpillar stage, but also sawflies, flies, mosquitoes, beetles, caddisflies, and even silverfish. Baculoviruses are unique among insect pathogens in having no close relatives infecting any higher animals or plants, making them very attractive candidates for biological control.

Polyhedra protect virus particles from the damaging effects of ultraviolet light in sunlight, so these viruses can remain active in the environment for many years. A susceptible insect may die after eating a plant contaminated by the remains of another insect that succumbed to the virus a year or more earlier. In the 1980s, Robert Granados and his colleagues at the Boyce Thompson Institute in New York worked out the sequence of events in infection of the alfalfa caterpillar, *Autographa californica*, by an NPV. When polyhedra reach the alkaline gut of the insect, the protective polyhedrin protein dissolves, releasing the virus particles, as was seen by Bolle almost a century earlier. The virus particles then fuse with the fingerlike microvilli on the surface of the caterpillar's midgut cells and are thus transported into the cell and finally into the nucleus. There the virus releases its DNA, which takes over the machinery of the midgut cell to make more virus. About twelve hours later, some rod-shaped nonoccluded viruses bud out through the gut cell membranes into the blood of the insect, which transports them throughout the body. The nonoccluded viruses proceed to infect many other cells; within twenty-four hours, polyhedral inclusions can be seen throughout the body of the insect. The caterpillar soon begins to show symptoms of illness as it stops feeding, changes color, and becomes lethargic. By the time the caterpillar dies several days later, its body has become a factory for the mass production of viruses, and the polyhedrin capsule protein may represent nearly 20 percent of all proteins in the body of the insect.

Many scientists had observed mass mortality of insect larvae caused by polyhedral viruses in the field, so it was natural for them to consider the possibility of using these viruses for insect control. The first targets were forest caterpillars. When forest caterpillars die from virus disease, their bodies literally melt down on the trees. Dead insects of this sort would turn customers away from an edible agricultural product such as an apple but did not pose a problem for the timber industry. As early as 1931, attempts were made to control a serious forest pest, the European nun moth, using polyhedral virus, but the results were disappointing. Then in 1944, R. E. Balch and F. T. Bird, at the Sault Ste. Marie Canadian laboratory, successfully controlled an introduced forest pest, the European spruce sawfly, using a virus. (A sawfly larva looks like a caterpillar, but in fact this insect is a relative of wasps.) Sawfly larvae proved to be highly susceptible to polyhedral viruses. Since Balch and Bird could not find useful viruses in Canadian sawflies, they imported diseased sawfly larvae from Sweden. The sawflies were rapidly killed, and the virus was transmitted through the insect's eggs as well as from one larva to another, resulting in long-lasting control of the pest.

Natural cycles of baculovirus-related deaths began to be observed in forest caterpillars. As scientists studied these boom-and-bust cycles, they found that when the insect numbers are low, the virus generally infects only a very small percentage of the hosts, but above a certain threshold population density, rapid spread and virus buildup from one generation to another leads to massive epidemics called epizootics. The quantity of virus in the environment then declines, owing to destruction through rainfall and sunlight, until an increasing insect population once again generates a virus buildup. These predictable cycles of insect outbreak, population crash owing to massive virus mortality, and slow resurgence of insect numbers occur repeatedly in forest ecosystems. Virus epizootics also occur in agricultural systems but are less predictable because of the much shorter life of the crops. Intensive study of these natural insect-virus cycles have led to important insights and mathematical models of disease that are applicable to disease epidemics in wild and domestic animals and in humans as well.

The potential for the use of baculoviruses for biological control was obvious, but several problems immediately became apparent. First, caterpillars continue to feed for several days before dying of polyhedral virus infections. While this allows the insect to build up large quantities of virus to infect the next generation, the farmer can lose part of his crop in the process. Just as with BT, viruses are restricted in the range of insect species that they can infect, so a single type of virus cannot be marketed as a general insecticide. And finally, viruses require living host cells to multiply and, unlike bacteria or fungi, cannot be produced in massive quantities on simple laboratory media. Nonetheless, imaginative scientists began to explore ways around these problems to produce commercial baculovirus products for insect control.

Edward Steinhaus and his students championed the idea of viral pesticides. The first marketable NPV product was initiated by Carlo Ignoffo, first at the U.S. Department of Agriculture laboratory in Texas and later at Bioferm, the same California company that perfected the first American BT product. Ignoffo's target was the corn earworm and its relatives. These caterpillars attack not only corn but

also cabbage, cotton, and other important agricultural crops, and
thus a product that could affect these pests would have a large potential market. The first NPV product, Viron H (later called Elcar by International Minerals and Chemicals Company, which purchased Bioferm), was registered for use by the U.S. Environmental Protection Agency in 1975.

Attention then turned to the Douglas-fir tussock moth, a notorious pest that was devastating forests from Canada to New Mexico. Douglas-fir tussock moth larvae are attractive caterpillars with brightly colored clumps of fur down the back. They tend to appear in huge numbers for about three years, and then abruptly disappear. Larvae were found hanging head down from the branches, slowly melting. Several types of virus particles were observed in these caterpillars by Steinhaus, Ken Hughes, and a young Swiss scientist, Mauro Martignoni, who had recently joined the Steinhaus laboratory.

In 1964 a program was begun at the newly established U.S. Department of Agriculture Forestry Sciences Laboratory in Corvallis, Oregon, to develop these viruses into a control method for this pest. Martignoni joined the Corvallis laboratory in 1965 and directed extensive research on the production, efficacy, and safety of the NPV of tussock moth. This research culminated, in 1976, in the registration of a product named TM BioControl-1, the first virus-containing insecticidal agent registered for use against a forest pest in the United States.

To produce the baculovirus insecticide for the tussock moth, special strains of caterpillars were selected that would thrive under laboratory conditions, and sterile, nutritious diets were developed to feed them. Larvae were fed a lethal dose of virus, and after they died they were frozen, homogenized into a slurry, centrifuged to concentrate the virus, freeze-dried, ground into a powder, and vacuum packed. When this virus-laden material was sprayed by airplane over many thousands of acres, the virus successfully controlled the tussock moth and set the stage for development of more baculovirus products. Currently, baculovirus insecticides are registered or

under development in the United States for use against fifteen different pest insects—including such serious pests as the gypsy moth—
in row crops, orchards, vineyards, and forests.

As Martignoni, Ignoffo, and others demonstrated, useful baculovirus insecticides could be produced using the insect itself, but the

Caterpillar killed by nuclear polyhedrosis virus

manufacture of these products was very difficult and expensive. Persons producing and handling these products may become allergic to them because they contain bits of insect cuticle or insect proteins. Perhaps the worst drawback is that viruses cannot be mass-produced on artificial laboratory media.

However, viruses can be cultured in living cells that have been persuaded to grow outside the body in liquid medium; these are called cell or tissue cultures. As early as 1904, cells from vertebrate animals, including humans, had been cultured on artificial media in laboratory dishes and bottles, but the culturing of cells from insects had been far from successful. Although William Trager had demonstrated the possibility of culturing insect cells in 1935, they rarely remained alive for more than a few days, and the original cells rarely divided, so they did not form a useful cell culture.

Thomas D. C. Grace, working at the Commonwealth Scientific, Technical, and Industrial Research Organization laboratory in Canberra, Australia, made the first major breakthroughs in insect cell culture. Grace reported that he had induced insect cells to multiply in culture in 1954, and in 1962 he reported the development of several continuously growing cell lines from a eucalyptus caterpillar. Grace used more than fifteen thousand eucalyptus caterpillar pupae to initiate these original cultures, which were the first long-lived insect cell lines. This breakthrough soon led to rapid progress in studies of insect viruses and insect physiology.

The critical component for the culturing of living cells outside the body is the liquid medium, which nourishes the cells and induces them to divide. Building upon the example of media developed for vertebrate cells, early efforts to develop insect cell culture media were based on chemical analysis of insect blood, or hemolymph. However, researchers eventually discovered that unlike the blood of vertebrates, which nourishes the cells, insect hemolymph also carries a heavy load of waste products and is therefore not a good basis for a culture medium. Fred Hink, a young professor at Ohio State University, discovered in 1970 that addition of fetal bovine serum, acquired from unborn calves during the slaughter process, along with some additional sources of vitamins greatly improved the growth of insect cells. Many other materials were eventually found to promote the growth of insect cells, and currently a number of media are commercially available to culture insect cells.

Even with a good medium, making a new insect cell culture is a real challenge. First, obtaining large quantities of the desired tissue requires the dissection of many insects (often, tiny embryos or early hatchlings) under sterile conditions, which is a time-consuming process. In addition, the medium must be acceptable to the cells in nutrient composition, pH, and osmotic pressure so that the cells will survive and divide and will not shrink or burst because of osmotic pressure changes. Moreover, sterility is critical, as bacteria, viruses, or other organisms from the original insect may contaminate the culture. Even after the initial culture is made, it requires continuous attention, careful medium changes, and great patience. With luck, after waiting for days or even weeks, the researcher may see cells beginning to grow out of the original tissue. As long as a year of care may pass before the culture becomes established and begins to grow rapidly. But once established, insect cells can be kept in culture for many years. More than four hundred cell lines have been produced from more than a hundred species of insects and their relatives, including many important pests. Large-scale production and long-term preservation of insect cells are now routine laboratory procedures.

Scientists immediately recognized the possible benefits of producing baculovirus insecticides in cell cultures instead of in insects. In 1970, Ron Goodwin and his colleagues were able to demonstrate efficient production of NPV in cell culture, with 100 percent of the cells infected. Because of the expense of fetal bovine serum, Goodwin and others developed serum-free media with known chemical composition, rapidly leading to large-scale production of NPV in cell culture. Many challenges needed to be overcome, however, in developing this important process.

One of these challenges lay in determining exactly how many viruses are in the product, since they cannot be easily counted under a microscope. The best way to count viruses is a system called a plaque assay. When a tiny sample of a virus product is placed on a

smooth, uniform layer of cells attached to the bottom of a dish, each
original virus particle should form a single hole, called a plaque, in
the cell layer when its progeny eventually destroy the cells. These
holes are counted and used to determine the amount of virus in the
original product. This technique was developed for mammalian cells
in 1952, but insect cells and viruses did not easily lend themselves to
plaque assay. In 1973 Fred Hink and his colleague Pat Vail discovered that an NPV from the alfalfa caterpillar, *Autographa californica*,
would infect *Trichoplusia ni* cells and form plaques. Hink and Vail
were then able to develop a very useful plaque assay system that
could be used to assess numbers of viruses, solving a major problem
in laboratory assays and commercial production of baculoviruses.

Many laboratories attempted large-scale culturing of insect viruses in cell culture with an eye toward making commercial products without using insects. This was easier said than done, however.
In the large glass or metal containers, called fermenters, used for the
large-scale production of microorganisms, cells would settle out or
stick to the sides. To get around this problem, the liquid medium is
agitated, and sterile air is added because the cells have a high re-
quirement for oxygen. But when air bubbles are forced into the culture liquid to add the needed oxygen, cells are often damaged as the
bubbles burst. And, of course, the entire system must be kept absolutely free of contaminating bacteria and other microorganisms. In
spite of all these problems, the culturing of insect cell lines came to
be accomplished in increasingly larger fermenters.

Perhaps the greatest problem arose through the rapid evolution
of the virus during continuous culture. Cell culture tends to select for
a higher percentage of nonoccluded viruses that pass from cell to cell
in tissue culture, rather than of viruses encased in polyhedral protein as are made in the insect. Unfortunately, nonoccluded viruses
are rapidly degraded when exposed to sunlight in the field. These
problems continue to plague large-scale fermenter production of insecticidal NPV products.

NPV insecticides used in the field suffer from some of the same
problems that limit the use of BT, as well as some that are unique
to these viruses. They have a limited host range, sometimes proving

active against only one or a few species, and unlike BT, NPV insecticides do not kill quickly. The viruses may take days or even weeks to complete their cycle in the insect, kill it, and reduce it to a pool of infectious virus. While the length of this process increases the likelihood that the virus will be passed along to more pests, it also means that the insects will continue to feed on and damage the crop or trees for several days before dying. These problems resulted in a limited market for the virus-containing products. However, developments in molecular biology were quickly used by imaginative scientists to solve some of these problems.

The 1970s were an exciting time for scientists working to decode and manipulate genes. Only twenty years after James Watson and Francis Crick had demonstrated the double-helix structure of DNA, the universal DNA code had been identified, and enzymes were commercially available that could be used to clip the DNA and paste it back together. From 1972 to 1978, several laboratories made rapid progress in the initial efforts to understand the genetics of baculoviruses. The plaque assay system developed by Hink and Vail permitted selection for mutants that were mapped to specific locations on the fragments formed when the DNA was clipped by enzymes. Finally, with the development of sophisticated techniques such as the polymerase chain-reaction machine and rapid-sequencing equipment, the entire DNA sequences of both the cabbage looper and silkworm NPV genomes were announced in 1994. These viruses were each found to contain more than one hundred different genes.

Once the genes governing the development of baculoviruses were known, efforts were focused on improving the speed with which these viruses killed their hosts. Several clever ideas resulted in genetically engineered viruses that were then tested as potential insecticides. These efforts focused on reducing feeding damage or interfering with normal development or behavior of the infected insect. Toxins from scorpions and an insect-parasitic mite were inserted into a baculovirus to significantly shorten the time that elapsed before death. These encouraging results led to the inser-

tion of additional genes, including hormones that control the molt-
ing cycle of the insect. These novel genetically engineered viruses
killed faster and reduced the damage to plants caused by the infected
larvae. The promising recombinant viruses were subjected to tests
to determine whether nontarget organisms could be harmed and
whether the viruses remained genetically stable if released into the
field. Tests revealed no evidence of deleterious effects, and in 1989
the first field trials began to evaluate the potential of recombinant
viruses for commercial use. Recombinant baculoviruses, as well as
the wild-type viruses already in use for insect control, are promising, safe, environmentally friendly insecticides with few deleterious
side effects.

While improving insecticidal activity of baculoviruses was an
obvious goal, Max Summers and his students at Texas A & M University had even more imaginative ideas for the potential utility of
these viruses. Summers knew that the polyhedrin protein consti-
tuted nearly 20 percent of the proteins in the insect body when NPV
completed its cycle. Indeed, the insect was literally converted into
an efficient polyhedrin factory. This suggested that the polyhedrin
gene must be under the control of a strong "promoter," the on-switch
gene that forces the massive overproduction of the polyhedrin protein. Summers realized that genes other than polyhedrin might be
expressed under the influence of this promoter if they were inserted
into NPVs and that this system might solve some of the problems
that had restricted the production of complex proteins in artificial
systems.

Thousands of medically important proteins have been produced
by inserting genes into bacteria such as *Escherichia coli*, but bacteria do not always produce cloned proteins that function as well as
we would hope. Bacteria do not have the proper systems to apply
sugars and other critical attachments to these proteins or to ensure
that the proteins fold into the correct three-dimensional structures.
These three-dimensional shapes must be perfect if the protein is to
function properly in cells of higher organisms. Insect cells, by contrast, do have the ability to properly form complex proteins that will

Max Summers

function in higher organisms, even in humans. Of course, these proteins could also be produced in mammalian cell cultures, but there is always the concern that potential disease organisms might be lurking in mammalian cells and contaminate the final product, whereas insect cells do not carry mammalian diseases.

The baculoviruses were found to have some very desirable characteristics for producing these valuable proteins. It was relatively easy to move new genes into NPVs by commonly used cut-and-paste methods. Special enzymes were used to snip the virus DNA, insert a new gene, and then patch the virus DNA back together. The virus was capable of carrying a lot of extra DNA, and cell lines were available for expression of these genes. In 1983, Gayle Smith, Mack Fraser, and Max Summers announced that they had removed parts of the polyhedrin gene, and in a second paper in the same year, they announced expression of a valuable human immune protein, beta-interferon, in place of polyhedrin using this system. Two years later, they demonstrated the expression of a second important human immune protein, interleukin 2, using recombinant baculovirus.

An explosion of interest rapidly followed in what became known as the baculovirus expression system. Thousands of genes were introduced into the viruses in place of the polyhedrin gene, and the system proved remarkably efficient at expressing foreign genes. It could produce not only simple proteins but also complex ones with multiple subunits, all assembled in the proper structure. The baculovirus expression system, in several different forms, has become a common laboratory tool that is sold by several companies in convenient, simple-to-use kits.

Perhaps the most exciting early product of this system was the first experimental vaccine for HIV, the virus responsible for the AIDS epidemic that has caused millions of human deaths. This vaccine was produced using recombinant baculovirus introduced into insect eggs and was tested in chimpanzees and humans beginning in 1987. At that time, scientists were optimistic that a vaccine against HIV would be produced and the problem quickly solved. Sadly, this has proven not to be true, and this vaccine, like many others since, was not successful.

An experimental vaccine for malaria was also constructed using this system, and it appeared promising in early tests beginning in 1999. Products for the diagnosis of and vaccination against a number of virus diseases—including swine fever, parvovirus of cats and dogs, Epstein-Barr virus (which causes mononucleosis), human papillomavirus (which causes cervical cancer in millions of women), hepatitis, and influenza—are all under development using this system. Many human and animal proteins with potential therapeutic or experimental uses have also been produced, and with the enormous amount of information coming from the Human Genome Project, this system will prove even more useful for studying these genes. Expression in the baculovirus system is efficient and permits production of valuable proteins for far less cost than extraction from human or animal tissues. Improved methods for expressing not one but many genes simultaneously have recently been developed. These methods will permit scientists to produce complex, properly expressed and folded proteins, similar to those produced in the human body.

Not only can genes be expressed in insect cell lines under the influence of genetically engineered baculoviruses, but they can also be expressed in infected insects. Susumu Maeda, collaborating with colleagues in California and Japan, produced dozens of important gene products using the NPV of silkworm and expressing them in live silkworms.

Early research had revealed that baculoviruses could enter mammalian cells and release DNA but would not multiply in these cells or kill them. With the increased ability to make recombinant baculoviruses, they have been used experimentally to insert "good" genes into cells to replace or overcome the effects of genetic disorders. In 1995 and 1998, baculoviruses were used to insert genes into human liver cells and, more recently, into central nervous system, artery, and pancreas cells. The cells responded by expressing a high level of the desirable gene product to potentially overcome genetic defects in the animal or human. This process, called gene therapy, is one of the most promising, and at the same time controversial, areas of modern medicine. Because baculoviruses cannot multiply in mam-

malian cells, they do not pose the potential hazards of the mammalian viruses that are currently used for this technique. Baculoviruses may soon play an important role in this medical advance as
well. Once again, a disease of insects has become important to the
health and welfare of humankind and may indeed be the virus that
cures.

5

Scourge of the South Pacific

The Rhinoceros Beetle

In 1909 the beautiful islands of Western Samoa received some un-invited visitors. A family of huge beetles arrived from Sri Lanka, perhaps on a ship. This beetle had a hard shell over its wings, and a sharp snout extending up and backwards from its head. This snout led to its being dubbed the rhinoceros beetle, *Oryctes rhinoceros*. When the beetles arrived on this lovely island, they immediately flew to the crown of a coconut palm, crawled down a frond stem into the crown, began to feed, and kept on feeding until they bored right into the heart of the tree. As the palm opened new fronds, each showed tattered signs of the chewing of the beetle. Palms are monocots, (that is, like corn, they develop from a single major crown or "heart" and do not produce multiple branches in the crown as trees such as oaks and maples do). This means that damage to the crown of a palm can eventually kill it. The arrival of the rhinoceros beetle was tragic for the Samoans, because coconut trees and their fruits have played a major role in the lives of South Pacific islanders for more than four thousand years.

The origin of the coconut palm (*Cocos nucifera*) is shrouded in mystery, but there is some evidence that it originated in northwest-

ern South America. Because a coconut can drift on the sea for
months and still germinate to produce a new palm, it could have floated all the way from South America to most of the South Pacific. The coconut palm can grow anywhere that is both warm and wet, as it cannot survive frost or drought.

Pacific islanders have been remarkably imaginative in finding uses for this plant. The wood is used for huts, carvings, and firewood; the fronds end up in clothing, furnishings, screens, and thatch; and the midribs of leaves become arrows, brooms, and even toothbrushes. The flesh of the nut is used for food or dried into a product called copra, which is extracted for oil. After the meat of the coconut is harvested, the shell and husk are used for rope, padding, charcoal, drinking cups, and many other important household functions.

The "milk" from mature fruits and the "water" from immature fruits are important sources of clean, healthy drinks. These products also have medicinal uses, which were documented as long ago as 1500 BC. According to folk medicine, coconuts were used to treat a long list of diseases from intestinal worms to tumors. Coco-

Rhinoceros beetle

nut water has minerals similar to those found in our blood and also contains sugars, some proteins, and vitamins. This water is used to hydrate infants with diarrhea and persons with cholera, and it is given as a tonic for the elderly and sick. Because the water of young green coconuts is sterile, it was used as a substitute for saline drips, saving the lives of injured servicemen fighting in the region in World War II.

Captain James Cook wrote admiringly of the coconut palms during his explorations of the South Pacific. Now, these valuable trees are also grown in South and Central America, Africa, parts of Asia, and India. Millions of people rely upon this plant for food and employment. Coconut oil, made from dried fruits, is used in foods, cosmetics, soap, lubricants, paint, plastics, margarine, and many other products. The United States alone imports around 200 million pounds of coconut oil annually.

Two more species, called oil palms—one from Africa (*Elaeis guineensis*) and one from the Americas (*Elaeis olifera*)—are also grown in many of the same countries for oil production. In Malaysia alone, oil palms cover thousands of hectares, and these palms have also become major crops in Indonesia, central Africa, the South Pacific, the Amazon basin, and Central America. Worldwide, more than 10 million hectares (25 million acres) are devoted to oil palm cultivation. Each hectare of mature palms produces fifteen to thirty tons of oil, which is converted into cooking oil and added to processed foods such as pizza, as well as being a component of soaps and detergents.

After the rhinoceros beetle arrived in 1909, it began to kill the palms of Western Samoa and then moved quickly to American Samoa in 1912. After a period of quiet, it arrived in Papua New Guinea in 1942, Fiji in 1952, Tonga in 1961, and the Tokelau Islands in 1963. Eventually the beetle invaded the entire coconut and palm oil–producing world, including Africa and the Philippines. In its wake came the destruction of the trees that many of the native people in these lands relied upon for their existence.

When the beetle burrows into the central fronds of a large mature coconut tree, it can first lower the production of coconuts. The openings made by the rhinoceros beetle then provide entry points

for infestations by another beetle (the palm weevil) and for fungal
diseases. Combined, these attacks often kill the tree. Young palms
are even more vulnerable and are rapidly killed when their growing
points are attacked. As a result, palm orchards are not replaced, and
production of oil in the region decreases.

The adult female *Oryctes* has two homes. She first burrows into
the palm heart to feed and then periodically flies out of the palm late
at night to find a place to mate and lay her eggs. She chooses dead
standing palm trees, decaying trunks, or heaps of compost, manure,
or sawdust as homes for the next generation of beetles. For an insect, the rhinoceros beetle lives a remarkably long time. The fat larva
feeds on decaying wood for more than six months, pupates, and then
lives for several months as an adult. During this adult phase, it can
fly long distances and injure many trees. Because these beetles are
secretive—flying only at night and hiding in the tops of tall trees—
it is hard to know when they have arrived and how many there are
until damage is done. These secretive behaviors and long life enabled

Left, Healthy coconut palm; *right*, palm damaged by *Oryctes* rhinoceros beetle

this insect to move long distances and cause significant damage to the coconut and oil palm industry.

As the rhinoceros beetle spread across the Pacific, scientists from several countries began to search for ways to control this devastating insect. Of course chemicals could be used to kill it, but these were expensive, and the countries with the problem were too poor to afford them. Also, the beetle's habit of hiding in the palm heart or in logs hampered the development of ways to get the chemicals to the target insect. Some plantations had partial success in controlling the beetle by cleaning up logs and sawdust piles where beetles bred. This was hard work, and the ability of the adult beetle to fly long distances meant that often a large area had to be cleared. A fungus, *Metarhizium anisopliae*, was known to kill the larvae, and this potential control agent was explored. Indian and Southeast Asian scientists cultured this fungus and attempted to use it to control the larvae. Then, in 1960, Paul Surany reported that he had found rhinoceros beetles suffering from several diseases in Southeast Asia and Africa. This study caught the eye of a young German scientist, Alois Huger, who decided to explore the possibility that there might be one or more pathogens that could successfully control this pest.

In 1963 Huger set out for the South Pacific in search of sick rhinoceros beetles. He explored North Borneo, Fiji, Western Samoa, and Malaysia (then called Malaya). In Malaysia he found beetle larvae that were bloated and translucent, and he also observed dying adult beetles. He passed this "Malay disease" to healthy larvae and found that they stopped feeding, suffered diarrhea, and became mottled with white spots. Sometimes their hindguts popped out of their bodies. He took specimens back to his laboratory in Darmstadt, Germany, where he used an electron microscope to examine them. Observing them under very high magnification, he found spherical and rod-shaped particles, especially in the nucleus of the fat body cells. These particles appeared to be arranged in crystal form.

Huger concluded that the crystalline particles were a virus, which he called rhabdionvirus, noting that it was different from the nuclear polyhedrosis and granulosis viruses that were known from

caterpillars. Later studies by other scientists found that this virus
had a peculiar DNA structure, double-stranded and supercoiled, enclosed in an enveloped rod-shaped capsule. Even more unusual was a tail-like structure protruding from one end of the rod. At first the virus was thought to be a baculovirus, like the caterpillar nuclear polyhedrosis viruses, but eventually it was placed in its own unique family, *Oryctes* virus, or ORV.

As Huger continued his studies, he discovered that the virus multiplies in huge quantities in the gut of the adult beetle. Although these beetles seem to behave normally, they excrete massive amounts of virus in their feces. This observation explains how the virus gets around; the adults infect the egg-laying sites when they lay eggs or gather for mating. In Huger's words, the adults become "flying virus reservoirs."

The possibility that this new virus might help the coconut plantations led another scientist, Karl Marschall, to attempt to introduce the virus into Western Samoa in 1967. Marschall placed chopped-up infected beetle larvae into artificial log heaps where he knew beetles would come and lay eggs and the larvae would feed. A year later, Marschall found that the virus had established itself and the beetle population was rapidly declining. But it was not clear whether the virus or some other factor was the cause of the decline. To investigate this question, Bernhard Zelazny went back to Samoa in 1970 and 1971 and found that only a small proportion of the insects were infected. The beetles were surveyed periodically during the 1970s, and the proportion of insects and breeding sites in Western Samoa infected with the virus fluctuated around 10 percent. More important, though, was the finding that the number of beetles had dropped by about half. Scientists were interested to find that more males than females were infected. It appeared that these males might be giving the disease to larvae when the adults flew to log piles to mate.

These results stimulated an Australian scientist, G. O. Bedford, to introduce the virus into Fiji and Wallis Island in 1970. Within four years, Bedford found that beetle damage to the trees had dropped from 90 percent to 20 percent overall, and in some areas there was no damage at all. Following this encouraging result, the virus was

released in log heaps and sawdust in Tongatapu in 1970–71, leading to a rapid drop in damage over the next year, and virus-infected
beetles were released into American Samoa in 1972, again leading
to a rapid drop in damage.

Since 1972, the virus has been introduced into many countries in
the South Pacific, Africa, and wherever the rhinoceros beetle has become a pest. These introductions often produced good results, especially when several different control techniques were undertaken
simultaneously. Dead trees were destroyed, or natural vegetation
was planted to cover the logs; rhinoceros beetles prefer to lay eggs
on logs out in the open, and covering them with vines reduced egg
laying and eventually lowered the damage to nearby trees. Spores
of the fungus *Metarhizium anisopliae* were also useful when they were
applied to sawdust heaps around sawmills. There they infected larvae and reduced the number of new adults.

As one final point of interest, it was learned that the adult beetles,
especially the males, are a critical component of maintaining the
virus and keeping the beetle population low. By flying from one
breeding site to another and defecating virus, they maintain constant virus infection in larval habitats. The small number of infected
adults remaining in plantations is therefore essential for keeping the
virus available to infect the next generation of larvae.

Bernhard Zelazny and his colleagues working in the Republic
of the Maldives off the coast of southern India discovered that the
Oryctes virus has another very important characteristic; it shortens
the life span of infected females. The adult female beetle does not
begin to lay eggs until about two months after she emerges from the
pupal stage and begins to fly around and feed on palms. The virus
reduces the adult beetle's life span by about half. This is enough time
for adult males and females to distribute virus but not enough time
to produce many eggs. The virus therefore takes advantage of this
unusual aspect of the beetle's life cycle to be transmitted from infected males and females to the offspring of uninfected females. Because the adults live a shorter time, the virus is also more efficient
in reducing beetle damage to the palms.

In recent years, the *Oryctes* virus has been introduced into India,

Tanzania, the Philippines, the Andaman Islands, the Arabian Peninsula, the Seychelles, the Caroline and Marshall islands, and many other regions of the tropical world. It continues to have an important role in controlling this damaging pest of oil and coconut palms. The virus persists even when the beetle populations contain as few as ten adults. In oil palm plantations of Malaysia, nearly 100 percent of the rhinoceros beetles are infected. And perhaps most encouraging of all, the rhinoceros beetle is still under control in Western Samoa, forty years after Karl Marschall introduced the virus into these beautiful islands.

6

Bug against Bug · Invertebrate Antibiotics

Invertebrates have survived for millennia in the presence of potentially lethal microorganisms. Even habitats seething with bacteria, such as the runoff from cattle farms or sewage treatment plants, are also teeming with invertebrate life. Obviously, these animals, like ourselves, have ways of fighting infection. This fact led scientists to ask, what allows animals as apparently simple as a water flea or a maggot to survive in such polluted environments? This question led to the discovery of chemicals with remarkable properties and potential usefulness to humankind.

The first scientist to ask these questions was Elie Metchnikoff. Metchnikoff was walking the beach of Messina, Italy, in 1882, enjoying a quiet day of observing marine animals while his wife and children were at the circus. Metchnikoff had been trained in invertebrate zoology in Russia and Germany, and he later became interested in marine biology while working at Naples, on the beautiful Italian coast. During his time in Naples, Metchnikoff became close friends with Alexandre Kovalevsky, a Russian scientist studying the embryology of marine invertebrates. Long discussions with Kovalevsky and hours of observations of these invertebrates led Metchnikoff to become fascinated with certain cells that he could observe circulating within their bodies. He saw particles being "swallowed" by these circulating cells. He thought that these "swallowing" or engulfing cells might be involved in two processes: digestion of food (these invertebrates had no digestive tracts) and defense against microorganisms.

On the beach at Messina that day, Metchnikoff collected a starfish larva—a delicate, transparent organism. On an impulse, he stuck a rose thorn in the larva and took it back to his room to observe its responses. The next day he was excited to see that cells had surrounded the thorn, apparently blocking the intruder from further injuring the starfish. These observations soon led Metchnikoff to study this phenomenon in other invertebrates, including insects and

Elie Metchnikoff
(1845–1916)

the water flea, genus *Daphnia*. He found that cells in *Daphnia* seemed to be attracted to yeast cells invading the body and then surrounded and swallowed them. His work on *Daphnia* convinced Metchnikoff that engulfment of invaders by circulating cells is critical to the defense of invertebrates. He coined the term *phagocyte* ("eating cell") for these defensive cells. In 1884 Metchnikoff published his theory of cellular immunity in invertebrates, the first demonstration of this process in any animal. He soon observed phagocytosis in many other animals, including mammals, and demonstrated that even pathogens as deadly as anthrax could be phagocytized by guinea pig cells.

In 1887 Metchnikoff became discouraged with the lack of respect for his work among his Russian colleagues, and he accepted the invitation of Louis Pasteur to join the group that was to become the Pasteur Institute in Paris. There he continued his studies on immunity, focusing on mammals. Toward the end of his life, however, he returned to the study of immunity in insects, concluding a long lifetime of discovery that led to a Nobel Prize in 1908. These discoveries ultimately led Metchnikoff to be regarded as the founder of the science of immunology. A bust of Metchnikoff, his long hair flowing, still adorns the immunology facility at the Pasteur Institute, and his ashes rest in a place of honor in the library.

The Pasteur Institute was an exciting place for scientists interested in basic questions of immunity around 1900. During his studies on the silkworm in southern France, Louis Pasteur had observed black spots surrounding invading fungal and protozoan pathogens and around wounds caused by scratches and bites. In 1919 Serge Metalnikov, another colleague of Kovalevsky's, joined the team at Pasteur Institute at the invitation of Emile Roux, who became director after the deaths of both Pasteur and Metchnikoff.

This international collaborative group asked several very important basic questions about insect immunity that can apply to humans as well. Pasteur Institute researchers wondered, for example, how can insects swallow bacteria that are highly pathogenic to humans and not become infected? The insect gut appeared to be a natural barrier to infection. A third Russian at the institute, Constantin Toumanoff (who later became engaged in important work on

Bacillus thuringiensis), proposed that digestive enzymes may destroy
these microorganisms. Then the scientists asked, how do insects resist microorganisms that do enter their bodies? Insects often incur injuries, as Pasteur had seen on silkworms. Also, parasitic wasps inject their eggs into the insect host, sometimes injecting bacteria as well. Metalnikov—by injecting caterpillars with a variety of bacteria, dyes, and stains—explored the mechanism of phagocytosis first observed by Metchnikoff. Metalnikov observed that the extent of phagocytosis varied depending on the material he had injected, and he noted that this cellular process seemed very complex. If bacteria were injected in large numbers, the protective cells appeared to cooperate to produce a large, black cellular mass. If the insect was lucky and the bacteria were not too pathogenic, the insect's cells would succeed in killing and digesting the bacteria, and the insect would survive. Metalnikov was convinced that the insect's survival depended on its phagocytic cells, and he showed that phagocytosis varied depending both on the insect and on the bacterium. For example, he found that the bacterium that causes tuberculosis was rapidly phagocytized and destroyed, while the one that causes leprosy was not. He also found that phagocytic cells could surround a large invader such as a wasp larva or seal up an injury, while producing a black pigment known as melanin. This explained both Pasteur's observations in silkworms and Metchnikoff's groundbreaking observation with the starfish larva.

The successful immunization of humans against anthrax and rabies by Pasteur in 1881 and 1885 were, of course, well known by scientists at the Pasteur Institute. Therefore, an obvious experiment was to attempt to immunize insects as well. Metalnikov, André Paillot, and several other scientists independently succeeded in immunizing insects by injecting or feeding them heat-inactivated or chemically inactivated bacteria. To their surprise, they found that insects could become immune to further challenge by the same bacteria in only one day! They were well aware that in contrast, immunity in mammals rises slowly over several weeks after vaccination. Immunized insects did not exhibit a very strong immune response and could still be killed by huge numbers of bacteria, but they were definitely less sensitive to challenge with live bacteria than their non-immunized counterparts were. How could this be?

For many years, these immunologists were divided into two schools of thought, and each group was quite adamant in its conviction about the process leading to insect immunity. One group, led by Metalnikov, was certain that insect immune defense was primarily cellular, as originally described by Metchnikoff. The other group, including Paillot and his colleagues, was equally certain that there was a noncellular, "humoral" component involved as well (our antibodies are an example of humoral immunity). At this time, the importance of antibodies in human immunology was just beginning to be understood.

Metalnikov immunized caterpillars, took a sample of their blood, and heated it to 65 degrees Celsius. This temperature should have destroyed any cellular activity, but Metalnikov found that the blood was still capable of destroying bacteria. Even though this experiment appeared to show that the immune activity was not totally attributable to cells, Metalnikov stubbornly continued to believe that cellular immunity was the principal immune response. Finally, however, an experiment with his colleague Vitali Chorine, published in 1930, convinced him of the importance of antibodies in insect immunity.

Meanwhile, another French group, including Paillot and a former student of Metalnikov's, Vladimir Zernoff, found heat-stable chemicals in the blood of immunized insects and called these chemicals bacteriolysins. Zernoff was even able to immunize a susceptible caterpillar by transferring blood from an immunized one, a process called passive immunity. The second caterpillar remained immune for nearly five days. These results convinced Zernoff and Paillot that some chemical or chemicals (and not just cells) in the blood of the caterpillar were involved in immunity. During this period of intense interest in immunity, French scientists used the immune system of insects and other invertebrates to gain important insights into the complex immune systems of all animals. Metalnikov wrote in 1924 that "immunity is a general fact common to all living beings. This

is why it is necessary to look for the causes and the explanations in
general biology."

With the retirement of many of the French scientists around 1930, the study of invertebrate immunity essentially stalled for nearly twenty-five years. In the mid 1950s, scientists again began to be interested in how it is possible for insects and other invertebrates to survive in an environment filled with potentially lethal micro-organisms. The assumption was always in the back of the minds of scientists that invertebrate immunity was merely a simplified version of vertebrate immunity. The term *antibody* was commonly used for immune materials in the blood of immunized insects. In his *Principles of Insect Pathology*, published in 1949, Edward Steinhaus commented on the unusual speed of immunization of insects, and the weakness of this immunity. But even he had no explanation for these remarkable differences from mammalian immunity. Then, in 1956, Jack Levin and Fred Bang observed clotting in the blood of bacteria-injected horseshoe crabs; this was the first in a series of new discoveries that would eventually lead to a major change in our understanding of how invertebrates defend themselves against diseases.

Two young graduate students were pursuing Ph.D.s in the mid 1950s using a series of vertebrate immunology techniques to investigate insect immunity. John Briggs was studying under Steinhaus in Berkeley, California, and June Stephens was pursuing a Ph.D. in Canada. Briggs injected several species of caterpillars with live or heat-treated bacteria. Periodically after inoculation, he cut a leg off the caterpillars and collected the blood. When this blood was mixed with live bacteria, it inhibited the ability of the bacteria to multiply on agar medium. The activity was found to be in the liquid as well as the cellular parts of the blood. As Metalnikov had found, this activity increased very rapidly (within the first twenty-four hours), and vaccinated caterpillars were later able to tolerate injections of bacteria without dying. However, Briggs also found antibacterial activity in the blood before immunization, although the activity was definitely increased in immunized insects. This preimmunization resistance is called innate immunity, while immunity arising after immunization is called acquired immunity. Insects apparently have both forms. Briggs also found that the immunity in the insect blood persisted after high-temperature treatment, something that usually destroys mammalian antibodies. He cautiously concluded that a new concept needed to be developed about insect antibodies.

In 1958, Fred Hink, working under Briggs as a graduate student, tried a new chemistry technique called gel chromatography to isolate antibacterial substances from immunized caterpillars. Using this technique, Hink and Briggs found two peaks of activity, which they called Factor A and Factor B. Only Factor A conferred passive immunity (that is, increased the immunity of caterpillars injected with the substance). Factor A was heat stable and was small —only about 7,000 daltons in molecular weight. By comparison, mammalian antibodies are between 136,000 and 1,000,000 daltons in molecular weight. Factor B was even smaller, about 2,000 daltons, and appeared only in the blood of immunized caterpillars, not in that of naïve caterpillars.

At the same time, Stephens was performing several similar experiments that she published in 1959. Her results agreed closely with those of Hink and Briggs. In 1962 she and her colleague J. H. Marshall also attempted to isolate the active principle from immunized insects. They found molecules that were small, heat stable, and resistant to degradation by certain enzymes and therefore were probably not proteins. Stephens agreed with Hink and Briggs—these were not like mammalian antibodies but were instead something else that had never before been described. The immune factors that Stephens was able to isolate did not respond the same way as mammalian antibodies to a number of chemicals. For example, the activity of most mammalian antibodies is destroyed by the enzyme trypsin, whereas the activity of the insect immune factors was not. Her work clearly illustrated that insect immunity differs in many ways from mammalian immunity.

In a review article in 1963, Stephens bluntly declared that insects do not have antibodies. She recognized that many unexplained questions remained: If acquired immunity was not attributable to antibodies, then what substance or substances could be producing it? How much of this immunity derived from cells and how much from humoral substances? And why were caterpillars resistant to injection with millions of some bacteria (for example, *Escherichia coli*, better known as *E. coli*), while only a few hundred cells of other species were lethal? Stephens concluded that "the surface is barely scratched" in this intriguing field of study.

Although several attempts were made in the 1960s to rise to Stephens' challenge, real progress was not made until the early 1970s. This advancement was partly because of the development of increasingly sophisticated chemical purification and analysis equipment and to rapid development of techniques in genetics. But equally important was the attraction of a young Swedish scientist to insect immunology.

In 1972 Hans Boman and his colleagues Ingrid Nilsson and Bertil Rasmuson performed a series of experiments in which they injected male fruit flies, *Drosophila melanogaster*, with three species of bacteria and then counted the bacteria that remained in the flies for several days after injection. They found that one of these bacteria, *Aerobacter cloacae*, was not lethal but in fact acted as a vaccine. They could use this bacterium to immunize the flies against later challenges by other, lethal bacterial species. The immune activity was not specific to any one type of bacteria and appeared to prolong the life of the fly. Blood from immunized flies inhibited bacterial growth on petri dishes as well.

Fruit flies are wonderful experimental animals, and experiments on them have led to many important discoveries in genetics. But these insects have one major problem: they are very small! Injecting them is a real challenge, and they yield very little blood for analysis. During one of those casual conversations that often take science in new directions, a distinguished colleague, Carroll Williams, suggested that Boman should attempt to immunize a much larger insect—the cecropia silk moth, *Hylophora cecropia*. In addition to their size, cecropia moths have another wonderful advantage; their large resting stages, or pupae, can be stored in a refrigerator for nearly a year, quietly waiting for the next experiment. When they are brought back to room temperature, they begin to metabolize normally, and they can provide plenty of blood for analysis. Cecropia pupae proved to be the perfect test-tube animal for Boman's lifelong fascination with insect immunity.

Boman repeated his *Drosophila* experiments with cecropia pupae and found that the pupae could be easily immunized by injection with nonpathogenic bacteria. These immunized insects rapidly eliminated injected pathogenic bacteria, while their unfortunate nonimmune counterparts developed billions of bacteria in their blood. The immune blood killed bacteria very rapidly; 10,000 *E. coli* were killed within five minutes after being incubated with immune blood. And again, the defense appeared to work against several unrelated species of bacteria.

Boman concluded from his initial experiments that the insects could raise defenses against both gram-positive and gram-negative bacteria but that these defenses were different. The categories "gram-negative" and "gram-positive" are derived from the way in which these bacteria respond to a set of stains developed by a microbiologist, Hans Christian Gram, in the 1800s. If the bacteria stain purple, they are gram-positive, but if they stain red, they are gram-negative. The difference in staining is because of major chemical differences in the cell walls of these groups of bacteria. Gram-positive bacteria include the rod-shaped *Bacillus* species (including *B. thuringiensis*), while gram-negative bacteria include many that are common in the environment and in our bodies, such as *E. coli*. Boman recognized that this was a system that could be used to investigate the chain of events leading to immunity in insects. In 1974 Boman referred to the insect immune process as "a hopefully unknown fish which we have on a hook somewhere out in the sea."

After many thousands of silkworm pupae had donated their lives to experiments in the Boman laboratory and elsewhere, several of the questions posed many years earlier by the scientists at the Pasteur Institute were answered. First, it was learned that the enzyme lysozyme is induced by immunization. This enzyme, common in egg whites and human saliva, destroys gram-negative bacteria and is responsible for some of the activity of insect blood. Interestingly, the lysozyme eventually isolated from the silk moth proved to be quite similar to the one in chicken eggs. However, many bacteria that are resistant to lysozyme are nevertheless destroyed by immunized insects, so researchers thought that perhaps lysozyme was not the whole story. As increasingly sophisticated chemical purification and analysis techniques were used to analyze immune serum, the silkworm pupa was discovered to have yet another wonderful characteristic: nearly all of the proteins detected after a cecropia pupa was immunized were those involved in immunity. In this resting state, the pupa turns on only its immunity genes in response to bacterial challenge, while the rest of its genes remain dormant. Purification and identification of these immune substances was therefore much easier without the presence of other confusing chemicals.

In 1979 Boman and his colleagues Dan Hultmark, H. Steiner, and Torgny Rasmusen succeeded in separating lysozyme from the blood of immunized silk moth pupae and discovered that they had also simultaneously isolated two more chemicals that inhibited growth of bacteria. They named these chemicals "cecropin A" and "cecropin B," in honor of the cecropia moth from which they were isolated. Eventually four more cecropins were discovered, leading to a total of six different but closely related molecules. The cecropins were found to be small, only around 4,000 daltons in molecular weight. Using a technique borrowed from protein biochemistry, the Swedish group subjected the blood of immunized moth pupae to gel electrophoresis, which separated the components according to their size. Then they overlaid the gel with a layer of agar containing bacteria and incubated the bacteria on the gel overnight. The bacteria grew to form a solid layer in the agar except in certain spots where the immune molecules had come to rest in the gel. Using this technique, researchers found cecropin-like molecules in many different species of moths, flies, and beetles. New techniques from molecular biology were used to isolate the genetic messages that were turned on during the immune response. These messages led to a sophisticated understanding of the genes involved in this process. This activity happens very fast, within two to four hours, which is why the insect immune response appears much faster than that of mammals.

The cecropins were found to differ in their activity against various kinds of bacteria. The A and B forms were highly active against both gram-negative and gram-positive bacteria, while others had a narrow range of activity only against certain gram-negative forms. *Bacillus thuringiensis*, the important biological control agent for pest caterpillars, was found to be totally resistant to cecropins. *Bacillus thuringiensis* manages to avoid attack by cecropins by having a resistant cell wall and by producing inhibitors of its own that interfere with the activity of the cecropin. Another bacterium, which is symbiotic with an insect-parasitic nematode, manages to avoid being killed because the nematode produces an inhibitor of cecropins. Thus, not only have insects developed ways to kill bacterial invaders, but also the invaders have found ways to bypass these safeguards.

In many insects, the role of bacteria is not invader but assistant. As discussed in a later chapter, many insects rely on symbiotic bacteria to provide essential components of their diet. Symbiotic bacteria may live in the gut of an insect such as a cockroach or termite where they assist in digestion, or they may reside completely within the cells of insects such as aphids and whiteflies, where they produce essential amino acids. It is critical to this symbiosis that the immune system of the insect "host" should not destroy the "guest" bacterium. The interactions of insects and bacteria are obviously very old and very complex indeed.

A much larger protein called P5 was found in the blood of immunized moth pupae by Boman and Albert Pye; when initially isolated, it appeared to have no antibacterial activity. However, when this large protein (20,000 to 30,000 daltons) was rediscovered in 1983 by Boman and Hultmark, it was found to actually be composed of up to six different proteins. These proteins, called attacins, were found to have a narrow range of antibacterial activity (against only *E. coli* and a few other bacteria) but in addition assisted the cecropins and lysozyme in attacking the outer membrane of invading bacteria. Thus these three groups of immune chemicals—lysozyme, cecropins, and attacins—work together like an army to destroy most invading bacteria. Lysozyme takes on gram-positive bacteria, cecropins kill both gram-positives and gram-negatives, and attacins destroy a few specific gram-negatives while helping the other two groups of molecules to do their job.

Within a few years after their discovery, the genes for cecropins and attacins were cloned, and their chemical structure and activity were understood. Cecropins were found to be similar in structure to mellitin, the main component in bee venom. But one very important characteristic of these molecules set them apart from mellitin and many other potent antibacterial agents—they did not damage mammalian cells. This discovery opened the door to the possibility of using them as antibiotics against human pathogens; they could kill invading disease-causing organisms without harming the infected person.

Meanwhile, small peptide antibiotics were being found in other animals as well, and some scientists even found similar chemicals in plants. Not surprisingly, in retrospect, animals that also live in habitats full of microorganisms (for example, frogs) were found to have a set of chemicals with a wide range of antimicrobial activities. These chemicals were found in the skin, gut, and respiratory tract secretions of frogs and pigs, and similar compounds were found in human skin. Not only are these molecules released free into the blood and secretions of the animals, but also many new ones were found attached to the surfaces or within the phagocytic cells first observed by Metchnikoff a century earlier. The dozens of antimicrobial peptides eventually discovered were found to fall into several different categories, and insect-derived chemicals appeared very similar to some found in higher animals, including humans. Our innate immune system is apparently rooted deep in our evolutionary past.

Many of these peptides were found to have remarkable activities that prompted further exploration of their potential to be formulated as drugs against human diseases. Most human-pathogenic bacteria, and particularly ones that infect hospitalized patients, have developed resistance to the antibiotics that are used to control them.

We tend to ask doctors to prescribe antibiotics without knowing whether these medications will kill the bacteria causing the problem or even whether bacteria are involved in the condition at all. We also feed large quantities of antibiotics to cattle and pigs to enhance their growth. This has led many dangerous bacterial pathogens to rapidly develop resistance to these antibiotics. In addition, bacteria can "trade" resistance genes, so that resistance in one species can lead to resistance in another species in the same person practically overnight.

But antibiotic-resistant bacteria cannot survive treatment with cecropins and other related peptides, apparently because of the mechanism used by the peptides to kill the bacteria. These peptides bind specifically to the surface of bacteria and poke holes in the cell walls, a mechanism that is different from the ways that currently available antibiotics work. This means that bacteria that have developed resistance to standard antibiotics, such as penicillin, are not resistant to cecropins.

The second important quality of cecropins is their lack of harmful activity against human cells, unlike some commercial antibiotics. Of course it is important for the insect that its cecropins do not attack its own cells, and mammalian cells are not that much different. Cecropins target molecules that are on the surface of bacteria but not on cells of insects or higher animals, including ourselves.

Finally, these peptides are very attractive because they kill pathogens very quickly and are not very specific in their targets. This means that they can stop a serious infection within a few hours and can be used for many different types of infections. Currently, bacterial diseases must often be diagnosed by culture and identification of the offending pathogen before an appropriate antibiotic can be prescribed. These steps could be bypassed with use of a broad-range, rapid-action, truly safe antimicrobial agent.

Although most early studies of insect immune peptides focused on peptide activity against bacteria, it soon became clear that many of these peptides were active against fungi, protozoa, and even nematode worms. The human yeast pathogen *Candida albicans* was found to be highly sensitive to cecropins. *Candida* is the cause of serious yeast infections associated with contact lenses and can be life-threatening in persons whose immune system has been compromised by chemotherapy or HIV. Pathogens causing three terrible diseases transmitted by insects—malaria, leishmaniasis, and elephantiasis—were also found to be killed or inactivated by cecropins. Since mosquitoes should be capable of making cecropins themselves, scientists wondered about the possibility of genetically "immunizing" the mosquito to prevent it from becoming a carrier or vector of the disease. These experiments are under way, aided by information from the complete genetic sequencing of both malaria and its principal vector, *Anopheles gambiae*, in 2003. Mosquitoes have been genetically modified to make them unable to vector some of these parasites. The challenge remains to replace the vectors with these genetically immunized mosquitoes in areas where the diseases are rampant.

Obviously, it is not possible to extract sufficient amounts of these peptides from insects to do many experiments. Genes for cecropins and attacins have been cloned and expressed in many systems, including human and other mammalian tissue culture cells and the baculovirus expression system. These genes have been altered and recombined in imaginative ways, such as combining a cecropin with a frog-skin peptide to produce highly specific, targeted activity. Many possible targets for cecropins have been proposed. These include treating septic shock, sexually transmitted chlamydia infections, and even bacterial pathogens of fish. Apple and pear trees have been transformed with attacin genes to provide resistance to bacterial diseases, including fire blight. Mice have been given cecropin genes to render them resistant to the spontaneous abortion pathogen brucellosis, a major problem in domesticated livestock. A modified cecropin gene was introduced into blood vessels of human heart patients and successfully reduced the narrowing of arteries after surgery. In the early 1990s, cecropins were discovered to have antitumor activity, a finding that has led to research on these potential anticancer agents. Many experiments are currently under way to explore these and other potential uses of these unique molecules.

If you ask a medical immunologist whether insects have an immune system, you will probably be told that they are too primitive to have antibodies or a complex immune system. And yet these animals have survived in the presence of pathogens for many millions of years. It is now clear that antimicrobial peptides developed by insects (and probably by their long-dead ancestors) have remained as components of immune systems throughout evolution. Insects are definitely not simple or primitive, and our rapidly developing appreciation of their ability to fight infection is leading to the creation of important products that may improve our lives.

7

Saved by a Crab

Limulus
Amoebocyte
Lysate
Assay

Insects are not the only invertebrates that have methods of fighting disease, as Metchnikoff observed with starfish in 1882. One of these invertebrates has donated a valuable tool to medical science. Two stories will illustrate why this tool is important.

The surgeon was pleased with Marie's progress after surgery. Her tumor had been removed, and it was not cancerous. Now she lay quietly in her bed in intensive care, with a tube in her arm dripping vital saline solution into her blood system to replace the fluid lost when she was in surgery. By the next morning, however, the intensive care nurse was anxiously watching as Marie's temperature inched steadily upward. She was trembling and sweating, miserable with fever. She was showing signs of shock, and her blood pressure was dropping dangerously. The surgeon and attending physicians bent over Marie with great concern: what had happened? Had she gotten an infection during surgery, or was some other problem causing her dangerous fever? Samples of her blood and the saline solution dripping into her arm were quickly sent off to the laboratory for analysis. Within a few hours, the results came back, showing that the saline solution had signs of the presence of fragments of bacteria; the medical team focused immediately on the tubes providing the fluids that were dripping into her arms. Her saline solution drip had been contaminated— not with live bacteria but with remnants that had slipped into the solution somewhere in the production line, many months earlier at a pharmaceutical company. With a change of tubing and solutions, Marie's fever began to drop, and she soon recovered from what was a potentially fatal error.

Susan complained to her general practitioner that she had had a fever on and off for weeks. He asked her whether she had any other symptoms, but he could not decide whether she had the flu, some other virus disease, or a more serious disorder. Finally, however, he asked whether she had experienced any injuries just before the fever started. Susan replied, "No, but I did have my wisdom tooth extracted." The doctor ordered a blood test and within a day knew that Susan had a bacterial infection hidden somewhere in her body. He suspected that the dental work was the culprit, and after careful inspection of her mouth, he found that, indeed, she had an infection deep in the bone where the wisdom tooth had once stood. The doctor drained the infected site and prescribed potent antibiotics. Her temperature dropped, and her jaw was soon healed.

In each of these two stories, the test that detected the presence of bacteria or their fragments originated in the blood of a crab.

More than two hundred years ago, physicians learned that injection of fluids nearly always led to fever. The German scientist T. Billroth suggested in 1862 that some substance present in the fluid was responsible for this fever, and called it a "pyrogen," although he did not know what it was. In 1874 P. L. Panum correctly proposed that the fever was caused by the presence of gram-negative bacteria.

Gram-negative bacteria are literally everywhere: on our bodies, on nearly all the surfaces we touch, in food, and even in low numbers in the water we drink. Many disease-causing bacteria, such as those responsible for strep infections and cholera, are gram-negative as well. Keeping gram-negative bacteria out of a drug during manufacturing or away from a medical device during production in a factory is very difficult. Although killing these bacteria with heat or chemicals is relatively easy, even steam sterilization of the devices or drugs will not completely remove traces of these bacteria.

The membranes of gram-negative bacteria have a component called a lipopolysaccharide (LPS), a fatty acid with a long-chain sugar attached. This is the molecule that induces fever. Even if the bacteria are no longer present, as in Marie's saline drip, the traces of LPS can cause dangerous fever. And if there is a hidden infection, such as in Susan's tooth, the LPS will enter the blood and lead to fever. Gram-negative bacteria can also accumulate in large numbers, called septic foci, in the urinary tract, lower respiratory system, bile system, or gut in addition to surgical devices or wounds. These foci release LPS into the blood. Although the body's response to LPS (rather than the LPS itself) is what is toxic, LPS is called a bacterial endotoxin. Endotoxin triggers many of the body's metabolic and defense systems, starting a cascade of events. These events take place over a period of several days, leading not only to fever, but also to aggregation of blood platelets, anaphylactic shock, liver damage, heart damage, and even death.

From the 1940s until the 1960s, LPS was detected using live rabbits. Suspicious samples would be injected into four to eight rabbits, and then the animals would be watched for an increase in temperature. Unfortunately, this test is not very sensitive to low levels of LPS, and gives only a positive or negative response; it cannot be used to detect the quantity of LPS present.

A better system for detection of LPS was needed. This new assay was developed based upon research done in the late 1800s and early 1900s, when curious scientists began to study the blood of invertebrates. W. H. Howell, Leo Loeb, and K. C. Blanchard studied the blood of the oldest arthropod in the sea, the horseshoe crab. These primitive armor-plated animals have a large horse-hoof-shaped body with a flat abdomen and a thin sharp tail, and they resemble their fossil ancestor, the trilobite. There are only four species still in existence, but these have been in the seas for around 360 million years—the horseshoe crab truly is a living fossil. Some horseshoe crabs grow up to two feet (0.6 meters) long and a foot (0.3 meters) across and can live for up to eighteen years. These large animals contain blood that is not red but blue, because the oxygen-carrying pigment contains copper instead of iron, and this blue pigment is in the liquid part of the blood rather than in cells, as in our blood. The crab's blood contains only a single type of cell, called an amoebocyte or granulocyte.

During their studies, Howell, Loeb, and Blanchard observed that when they exposed the blood of the horseshoe crab to gram-negative bacteria, the cells burst and the blood coagulated into a jellylike blob. Later studies showed that the amoebocytes, not the liquid part of the blood, were producing this response. The part of the bacteria that stimulated the cells to lyse and produce the jellylike coagulate turned out to be the endotoxin LPS. When crabs are wounded in the sea, gram-negative bacteria will very likely enter the wound, as these bacteria are present in high numbers in much of the ocean. Bacteria are trapped in the jellylike coagulate and are rapidly killed. The crab uses the presence of bacteria, and the LPS on their surface, to trigger the clotting response that may save its life.

In 1956 Fred Bang observed clumping and clotting of cells around bacteria injected into the Atlantic horseshoe crab, *Limulus polyphemus*. In collaboration with Jack Levin, Bang collected the cells from the blood, separated them from the serum, and lysed them. Bang and Levin separated the proteins from the cells into two fractions. They found that incubating samples of LPS with the first fraction and then adding the second fraction led to rapid formation of a gel. Japanese scientists working with a Pacific species, *Tacypleus tridentatus*, found that this gel is formed by a cascade of events leading to conversion of a soluble protein from the cells into a fibrous, gel-forming protein appropriately called coagulin. The LPS activates an enzyme that sets this chain of events into motion. Levin and Bang, followed by many other scientists, developed this system into what became known as the limulus amebocyte lysate assay, or LAL assay. In its simplest form, a small quantity of the sample (e.g., a pharmaceutical product or blood) is mixed with an equal quantity of extracts from lysed *Limulus* cells. The solution is incubated for one hour, and then the tube is carefully inverted. If the gel stays in the bottom of the tube, the test is positive. This system can detect amounts of LPS as small as one picogram (one trillionth of a gram).

In the 1970s, Stanley Watson, working at the Woods Hole Oceanographic Institute, developed a product based on the LAL assay for use in his research on marine bacteria. Pharmaceutical companies became very interested in this assay as a test for product contaminants, and Watson set up the first small company, Associates of Cape Cod, to manufacture the product. In 1973 the U.S. Food and Drug Administration approved the licensing of the first products. By 1983 the LAL began to replace the rabbit test for products for injection and inhalation. These products are labeled "nonpyrogenic, USP," stating that their safety has been assured using a test published in the *U.S. Pharmacopeia*, the rule-book for pharmaceutical products. The assay has also been approved for use in Europe and Japan because this assay is quick and can be used to accurately quantify the amount of LPS present in the sample. This allows the company making a product to trace any air, water, or surface contamination back to its source. Among the products currently being tested by LAL are dialysis materials, radiolabelled materials for medical scans, anesthetics, antibiotics, and even foods including milk. This system is absolutely critical to modern technological development of safe medical devices and drugs.

There is no question that the LAL system is medically useful. An environmentally aware person could ask, however, what the impact of this assay has been on the horseshoe crab. The valuable blood is drawn from crabs collected only once a year when they crawl onto the beaches to spawn. After their blood donation, the animals are released back into the ocean, presumably without harm. Obviously, it is in the interest of the companies producing the LAL assay to maintain good numbers of the crabs. Although no more than 10 percent of the crabs die after bleeding, there is evidence that Atlantic *Limulus* numbers are steadily declining. On the Atlantic coast of North America, horseshoe crabs are harvested in large numbers for use as bait for conch, a large snail that is a delicacy in China. Harvests of *Limulus* for bait have been reduced by U.S. government edict out of concern not only for the crab, but also for shorebirds that rely on crab eggs for food during migration. In Japan, *Tacypleus* is already considered endangered largely owing to overcollection.

To get around the problem of declining horseshoe crab numbers, scientists have begun to clone the genes from crabs to produce the proteins involved in the cascade of events leading to LAL gel formation. Their goal is to replace the crab with a system produced in the laboratory. Recently, scientists in Singapore have cloned the gene for the enzyme that initiates the cascade into caterpillar cells, using a modification of the baculovirus expression system. The enzyme continued to be secreted after thirty passages in cell culture. If the entire cascade of proteins can be produced in the baculovirus system, eliminating the need to harvest blood from the crab, two invertebrate pathology systems will have come together for the benefit of both humans and the humble but ecologically important horseshoe crab.

The LAL assay is only one of several important medical uses that humans have found for the horseshoe crab. A new test for fungal infections using a crab product is already in use in Japan. Potentially useful antiviral and anticancer proteins have also been found in the blood of horseshoe crabs. The skeleton of these crabs has been used to produce chitin to coat sutures and wound dressings, greatly reducing healing time. Horseshoe crabs were used in groundbreaking studies in 1926 that led to our understanding of the functioning of the human eye, and these studies resulted in Nobel Prizes for the scientists. Who can predict what other important medical discoveries will come from these valuable living fossils!

8

Sea Sickness

Diseases of Edible Shellfish

Many of us have had the unpleasant experience of becoming very ill a few hours after enjoying a meal of seafood. If the shrimp, mussels, or oysters that we feasted on are not cooked or handled properly, or if they come from a contaminated source, we will suffer from nasty diarrhea and vomiting. Ironically, the shellfish, too, can suffer from serious, often lethal, diseases.

Since prehistoric times, humans have been eating almost any creature that we can catch. Shell middens (large collections of shells discarded by prehistoric people) are often found near streams, rivers, bays, and estuaries, evidence of the longstanding human appreciation of these molluscs. In the past, crayfish, shrimp, crabs, clams, oysters, and lobsters were caught in rivers or oceans through the use of nets or traps. However, within recent decades, enterprising seafood producers have discovered clever ways to culture some of these animals, leading to a flourishing industry. Now, many of the shrimp or oysters in your local grocery store are cultured, not caught in the wild.

Let's first explore how shrimp are "farmed," a process called aquaculture or mariculture. Most cultured shrimp belong to one of several species in the family Penaeidae, which lend themselves readily to aquaculture. Initially, the most commonly cultured species were *Penaeus vannamei* in the western hemisphere and *Penaeus monodon* in the eastern hemisphere. However, in recent years, *Penaeus vannamei* became the dominant species in Asia as well when disease-free, improved genetic lines from the American Pacific region became commercially available.

Penaeid shrimp go through several different stages during their development. The adults mate, and the female lays eggs in the ocean or in a hatchery, always at night. Within one day, the eggs hatch into nauplius larvae, which resemble tiny spiders. These nauplius larvae are planktonic (that is, they float in the water, feeding on their egg-yolk reserves). Two days later, they change into another stage called a zoea, which has a long body and feathery appendages and feeds on algae and microscopic organisms. The zoea grows rapidly, and three or four days later, it transforms again into a mysis, which begins to look more like a shrimp. Finally, in another three or four days, it reaches a stage called a postlarva, which is easily identified as a shrimp. In the wild, the postlarvae migrate into estuaries, where rivers reach the sea. There they seek out salt marshes or mangrove swamps in partially salty brackish water and continue to grow into juvenile or subadult stages. In nature, these subadults migrate back into the sea, where they mature into adults to complete the life cycle.

Wild-caught or cultured adult shrimp are transferred to aquaculture tanks for mating. Each adult female shrimp can lay 50,000 to 1,000,000 eggs, which hatch into larvae that are then sold for commercial production, at a significant price. The larvae are grown up in ponds, where they feed on natural algae, brine shrimp, or processed food.

Artificial culture of shrimp began in the early 1970s, and now more than fifty countries participate in this business. Thailand is the largest shrimp producer, with Vietnam, Indonesia, China, India, and several other Asian countries not far behind. In the Americas, Brazil, Ecuador, Mexico, and several other countries in Central and South America also have large industries, as do Madagascar, Mozambique, Saudi Arabia, and Iran in the Indian Ocean and Middle East. In

Asia, culture may take place in small single-family farms, where

the ponds are often in or near coastal mangrove swamps. In larger, semi-intensive ponds, shrimp are fed commercial feeds. In large-scale intensive farming, shrimp are held at high density in sophisticated nurseries where water is pumped in and aerated, shrimp are fed nutritionally complete food, and wastes are managed. The most intensive culture of shrimp takes place in Thailand, where farms can cost millions of dollars to set up and turn out thousands of tons of shrimp each year. Shrimp culture in mangrove swamps has been strongly criticized for destroying the ecology of the tidal and mangrove areas and has been banned in many countries and significantly decreased in others.

When animals or plants are moved around the world by humans and massive numbers are confined to a relatively small space, a common result is the outbreak and spread of disease. In intensive mariculture, shrimp are fed commercial products that include fish meal and by-products of fisheries, including bits of crustaceans (such as the shrimp themselves) that potentially harbor disease. Importation of live shrimp and shrimp products from other regions of the world has also contributed significantly to the movement of diseases. Other factors—including pollution from local farms, increased water temperatures caused by El Niño occurrences, hurricanes, droughts, monsoons, and tsunamis—can impact diseases in shrimp fisheries as well. In some areas of the world, shrimp farms are washed away by storms about every ten years.

In wild and commercially reared shrimp, all types of disease microorganisms have been found, including bacteria, fungi, and protozoa, but the most harmful are virus diseases. We have discovered these viruses only in recent years, as interest in aquaculture has increased. The first confirmation of a virus disease in a marine crustacean was made by the French scientist Constantin Vago in 1966 in a shore crab on the Mediterranean coast of France. In 1971, Fred Bang (whose earlier research led to the development of the limulus amoebocyte lysate assay) found spherical viruslike particles associated with an infection in a European shore crab and showed that this infection could be transmitted by injecting blood from an infected crab into a healthy one. Bang proposed that the disease was "probably viral." Now, more than thirty shrimp and crab viruses are known. Because local shrimp species tend to be more resistant to local diseases, the movement of shrimp around the world for breeding has led to Western diseases being moved to the East, and Eastern diseases being moved to the West, often with devastating results.

John Couch, working at the U.S. Environmental Protection Agency laboratory in Florida, first described a virus disease in pink penaeid shrimp from the Gulf of Mexico in 1974. Couch observed pyramid-shaped glittering bodies in the cells of an organ called the hepatopancreas, and he realized that these must be virus polyhedra. The presence of polyhedra suggested that this disease was caused by a baculovirus, a group of viruses that had previously been seen only in insects. Since then, viruses in many different taxonomic groups have been found in shrimp. Among the most serious diseases for cultured shrimp are white spot syndrome virus, yellow head virus, infectious hypodermal and hematopoietic necrosis virus, and Taura syndrome virus.

White spot syndrome virus (WSSV) first appeared in northeastern Asia in 1992. It turned up in North America in a shrimp farm near Brownsville, Texas, in 1995 and by 1998 had spread to shrimp collected in the wild. The Texas shrimp industry is the largest in the United States, producing millions of dollars worth of shrimp each year and employing many people, so the sudden appearance of a disease was taken very seriously, and scientists were recruited to study this potentially devastating disease. Eventually at least five closely related viruses were found to cause similar symptoms in shrimp, crayfish, lobsters, and crabs. These viruses also appear to be related to baculoviruses (being similarly rod-shaped), although some have odd tail-like appendages at one end. They multiply in the nucleus of infected cells and cause the shrimp to develop a reddish appearance and characteristic white spots on their shells. WSSV is highly virulent to many shrimp and crab species, causing symptoms within only three to five days after exposure. Epidemics of WSSV spread very rapidly throughout countries cultivating shrimp in Asia and the Americas, leading to sudden losses of entire crops. It was found to be spread by contaminated water, decomposing infected animals, and cannibalistic behavior of the shrimp. The virus was moved from pond to pond by equipment and especially by importation of live shrimp for aquaculture and frozen shrimp products that are reprocessed for marketing in the United States. In impoverished areas of Latin America, WSSV has had a serious impact on local economies. In Ecuador, for example, nearly 80 percent of the country's production was wiped out in 1999 because of importation of infected live shrimp.

A second viral disease, yellow head virus (YHV), appeared in Thailand in 1992. Scientists found rod-shaped viruses inside the cells of infected shrimp. At first YHV was also thought to be a baculovirus, but studies soon showed that its genetic material was RNA, not DNA as in baculoviruses. YHV and several closely related viruses appear to belong to a new genus of viruses called *Okavirus*. Yellow head disease caused up to 100 percent mortality in giant black tiger shrimp, a valuable Asian species, within three to five days of the first appearance of symptoms. Infected shrimp first begin to feed ravenously and then they stop feeding altogether, turn pale yellow, and swell up, and their gills become white to yellow-brown. YHV is confined to Asia, at least at the present, although a similar virus has been found in Australia. North American shrimp are highly susceptible to YHV in laboratory experiments, so if it ever enters the Western Hemisphere, it could have devastating results.

A third virus, infectious hypodermal and hematopoietic necrosis virus (IHHNV), was first seen in 1981 in Pacific blue shrimp, *Penaeus stylirostris*, at an experimental shrimp farm in Hawaii run by the University of Arizona. Animals infected with this disease display bizarre behavior, which includes swimming erratically, rolling over and sometimes swimming upside down, and moving up and down in the tank. The infected shrimp develop white or buff color and sometimes small spots on their external skeletons. Eventually the victims fall to the bottom and die. In shrimp stocks that originated in Mexico and Central America, sudden epidemics led to the death of up to 90 percent of the cultured shrimp. Donald Lightner and colleagues from the University of Arizona eventually found that the disease was caused by a virus that produced odd inclusions in the cells. IHHNV was finally identified as a parvovirus, a distant relative of viruses causing disease in dogs and cats. Lightner and others also found that while the Pacific blue shrimp is highly susceptible to IHHNV, *Penaeus vannamei*, the shrimp cultured in much of the western hemisphere, is far less susceptible. However, *Penaeus vannamei* can become chronically infected and become stunted and deformed, which makes them unattractive for market. Once again, the culprit appears to have been brood stock that brought the disease into the rearing facility, although there are suspicions that birds may also spread the disease. IHHNV was eventually found in many other species of *Penaeus*, and it was probably introduced into the Americas from Asia during the 1970s, when shrimp culture was first begun.

In 1992, shrimp in farms near the mouth of the Taura River in Ecuador began to display pale reddish color with scattered black spots, fanned their tails, and eventually died when they attempted to shed their skins to grow. The shrimp cultured in Ecuador were *Penaeus vannamei*, and although these were fairly resistant to IHHNV, they were highly susceptible to what became known as Taura syndrome virus (TSV). Up to 90 percent of juvenile shrimp died in the first year, and Ecuador lost a crop worth nearly $400 million in a single year. Shrimp farmers in Ecuador, reluctant to admit that there was a virus in their stocks, did not take emergency cleanup measures in time to stop the spread of the virus. The disease moved rapidly to other shrimp-farming areas in Latin America, where it led to losses of more than $1 billion. It made its way to Hawaii in May 1994 and killed up to 95 percent of the crop there, before moving on to Texas in 1995 and South Carolina in 1996. This virus has the ability to be carried in the gut of flying insects and seagulls, which can rapidly move the disease from pond to pond and even from one country to another. Once again, Lightner and his colleagues discovered an unusual virus in TSV-infected shrimp. This time the genetic material was found to be single-stranded RNA, and its shape and replication led to its eventually being assigned to another new virus family, Dicistroviridae, distantly related to diseases in insects, fish, and birds.

Several tools have been developed to fight these virus diseases in shrimp mariculture. An important first step is assuring the growers that brood stock and larval shrimp are free of disease before they are brought into the ponds. Tons of shrimp are imported from Asia to North America each year; some of these are processed within the United States near coastal breeding sites for native shrimp, while others are sold as bait for anglers. Special concern has been raised about the possible importation of white spot and yellow head viruses, which are common in Asia. Techniques—including histology using microscope slides, antibody tests for the individual diseases, and, most recently, genetic tests using the polymerase chain-reaction machine—have been developed to search for viruses. Resistant strains are being selectively bred, and fisheries have instituted careful sanitation procedures to reduce the possibility of disease. Constant vigilance will be needed if we are to continue to enjoy an abundant supply of delicious, relatively inexpensive shrimp on our tables.

Shrimp are not the only crustaceans that we enjoy eating. Their relatives—crayfish, crabs, and lobsters—have also been on our plates for many years. Crayfish, a freshwater relative of shrimp, is a delicacy in many countries, and so when massive mortality of these animals was observed in Europe in the late 1800s, it caught the eye of scientists. The disease, called krebspest, or crayfish plague, was first seen in Lombardy, Italy, in 1860. It had moved to France by 1900, into Russia and Scandinavia by 1907, and finally to England in 1981. Crayfish are now rare in much of Europe, and some populations have been totally exterminated by this disease. In 1903, a German scientist, F. Schikora, found that crayfish plague was caused by an aquatic fungus called an oomycete, and he named this fungus *Aphanomyces astaci*. Oomycete fungi produce an unusual life stage called a zoospore, a cell with flagella; this zoospore swims around freely in the water searching for a crayfish.

Torgny Unestam and his Swedish colleagues have studied many aspects of the way in which this devastating disease kills its host. The zoospore attaches to the cuticle of the crayfish and sends out long, thin hyphae that burrow into the body of the animal at the soft regions between the segments and in the joints. It takes over the body and kills the crayfish within one to two weeks, and then the hyphae burst back out of the cuticle to send out more zoospores. Sick crayfish lose their normal fear of light, appear in open water in the daylight, and may be unable to move normally or right themselves if they are turned over. All infected European crayfish inevitably die. Unestam, David Aldermann in England, and other scientists eventually learned that North American crayfish are resistant to this disease, apparently because of their ability to raise an immune response that forms a black mass around the invading hyphae, slowly smothering the fungus. This disease was probably moved to Italy with imported North American crayfish in the late 1800s; once again, movement of animals from one continent to another resulted in massive disease outbreaks.

One of the most delicious and expensive seafoods is crab, which often appears as crab legs, crab cakes, or soft-shell crab on restaurant menus. Crabs are also crustaceans, relatives of shrimp. In order to grow, the crab, like other arthropods, must shed its shell and grow while the new shell is soft. Male Atlantic crabs seek out a freshly molted female and carry her around until she produces eggs, which he then fertilizes. When her shell has finally hardened, the female crab carries thousands of eggs glued to her abdomen until they hatch into free-living larvae. The larvae float in the sea and feed for several months while growing rapidly and undergoing multiple stages, similar to those in shrimp. Finally the young crabs settle to the bottom, where they spend the rest of their lives. Tons of crabs of many species, worth millions of dollars, are trapped for sale from both Atlantic and Pacific coastal areas each year. Trapping alone has reduced the crab populations in many regions, but disease has also had a devastating effect. In contrast to the situation with shrimp, for which viruses are the major culprits in disease, the worst diseases of crabs appear to be caused by tiny parasites.

Along the U.S. Atlantic and Gulf coasts, the blue crab, *Callinectes sapidus*, is the most important commercial species, generally ending up in crab cakes. In recent years, however, blue crab populations have crashed, leading to the bankruptcy of crab suppliers

and a marked increase in the cost of the product. In 1931 a parasite called a dinoflagellate was reported in the blood of crabs collected on the French coast by E. Chatton and R. Poisson, who considered it a rare disease. Chatton and Poisson named the pathogen
Hematodinium perezi. The disease remained hidden until 1975, when
it turned up in blue crabs on the U.S. coast, causing widespread disease. Now it is present all along the Atlantic coast and has recently
been found in infected crabs in Britain as well. *Hematodinium* infects
smaller crabs, especially in the fall and in areas where salinity is
high. Because high salinity encourages this disease, it is more likely
to break out in years when there is a drought and dies down during heavy rainfall. It often seems to completely disappear during the
winter but then increases as the temperature goes up in the spring
and summer and kills the crabs in the fall.

Dinoflagellates are a group of microscopic protozoa-like organisms that swim by means of two long flagella, one of which wraps
around their center in a groove. Many dinoflagellates are parasites
of both freshwater and marine invertebrates. Related organisms
produce lethal toxins including the fish killer *Pfisteria* and "red tide,"
which turns the ocean red and can sicken a person who eats a shellfish containing this flagellate.

Hematodinium has resisted efforts to culture it in the laboratory,
and so its life history remains somewhat unclear. We do know that
it has free-swimming forms that probably attack the crab and somehow enter its body. It then proliferates in eggs, as well as in stom-
ach and other tissues, and it consumes the disease-fighting blood
cells and the blood pigment that carries oxygen. Eventually the crab
smothers to death for lack of oxygen. Spores of the parasite then
leave the crab from its gills and hide in the surrounding water until
they contact the next crab. Crabs are quick to cannibalize a sick or
dead neighbor, a behavior that also leads to rapid spread of the disease. *Hematodinium* has led to major declines in the blue crab population along the U.S. Atlantic coast and is a threat to other crab
fisheries in Alaska, Australia, and Europe.

Hematodinium is only one of many protozoan diseases that cause
the death of valuable crabs. Others include bitter crab disease, which

causes the crab on your plate to taste rather like aspirin; pepper spot disease, which is carried to the crab via snails and parasitic worms; pink crab disease, discovered along the British coast and apparently related to *Hematodinium*; and cotton crab disease, which makes the meat look and taste like dirty cotton. These diseases have had serious effects on the profitability of local crab fisheries and have reduced the populations of crabs.

Our other favorite tasty crustacean, the lobster, has also not escaped disease. A bacterial disease, called red tail or gaffkemia, was first observed in Maine in 1946 and was originally named *Gaffkya homari*, the source of the term *gaffkemia*. After several name changes, the tiny spherical bacterium has finally come to be called *Aerococcus homari*. *Aerococcus* invades the heart and blood of the lobster and causes the animal to hemorrhage. Often the lobster looks normal until near death, but its blood is pinkish and cannot clot. Phyllis Johnson, a scientist at a U.S. East Coast marine laboratory, found that this bacterium was a highly virulent pathogen. Johnson discovered that if she injected as few as ten bacteria into a lobster, the bacteria would multiply rapidly and apparently were not inhibited by the immune system of the animal. The lobster would inevitably die within six to fourteen days.

Lobsters are not friendly creatures, and they frequently fight, especially when they are crowded in lobster pots or captive holding pens while waiting to be shipped to market. Lobster suppliers often force wooden pegs into the large claws of lobsters to prevent fighting, a practice that simultaneously forms wounds. *Aerococcus* finds its way into the lobster in wounds, and when the diseased lobster falls sick or dies, it is cannibalized by its neighbors, thus spreading the disease.

A second group of invertebrates that frequently finds its way onto our tables is hard-shelled molluscs, oysters, clams, and mussels. Oysters are intensively farmed in warm waters along the coasts of the United States and Australia. In Louisiana, oyster hatcheries are located where water quality is good and salinity is lower than in the open sea; these areas produce more algae, which serves as food for the oysters that grow in the waters. Adult male and female oysters are placed in shallow cold water for about a month and are then slowly warmed up to trick them into thinking it is springtime. The males then release sperm, which females sense and release millions of eggs, turning the water cloudy. Fertilized eggs hatch into a ciliated swimming stage called a trochophore, which eats algae and other floating plankton for about two days, after which it changes into a different stage, a veliger, which has a special organ that helps it to swim and feed. The veliger eventually grows a footlike structure and attaches to a solid object on the ocean bottom, where it grows into a tiny oyster called a spat. The oyster goes through all these stages within about two weeks after hatching. Oyster farmers provide pellets of oyster shells and other hard materials to lure the spat to attach, and then the spat are cultivated for another four to six weeks, after which they are placed into mesh bags on racks or nets off the bottom of the bay. There they remain, feeding on algae and other plankton, for another year until they are harvested.

Oysters have been farmed in New South Wales and Queensland, Australia, since 1870. In Queensland, three species of oysters are reared using spat collected from the wild on specially coated substrates used to lure the oysters to attach. These oysters start life as males and then change to females as they grow. The collected spat are laboriously culled by hand to make space for each oyster, since big, perfectly shaped oysters bring the best prices.

Oysters are keystone species in some marine ecosystems, converting organic material to energy. They are very efficient filter feeders; each oyster can swallow and clean forty to fifty gallons of water in a single day. They reduce algae outbreaks that can cause a marked drop in oxygen content in the ocean, leading to the death of many other organisms, such as fish, that rely on the oxygen.

Oysters have many enemies. Sitting glued to a solid substrate, unable to move, they are prey for crabs, starfish, boring sponges and mud worms, snails that drill into their shells, flatworms that attach to their surfaces, and barnacles and mussels that compete for space. But they also suffer from many diseases, and like crabs, their most serious diseases are caused by parasites.

During the early 1940s and 1950s, the catch of oysters along the U.S. Gulf coast and in Chesapeake Bay declined by nearly 90 percent. In 1950, a disease of American oysters was described by John Mackin and his colleagues in Louisiana oysters, a disease that caused massive mortality in this valuable fishery. They named the parasite *Dermocystidium marinus*. Frank Perkins, working at the Virginia Institute of Marine Biology, studied many aspects of this disease in the early 1970s, and eventually *Dermocystidium* was found in more than sixty species of molluscs from North America, Europe, Australia, Mexico, and the Caribbean. The disease breaks out when temperatures and salinity are at their highest. It moves from one oyster to the next by means of a flagellated zoospore, which penetrates the gills or gut of the oyster and then enters the body. It is carried through the body by blood cells and causes extensive damage before finally killing the oyster. The taxonomy of this complicated group of parasites has changed repeatedly over the years, and in 1988 the name was changed to *Perkinsus marinus*, in honor of the extensive work of Frank Perkins, although it is still often referred to as "Dermo." *Perkinsus* now is found along the North American Atlantic and Gulf coasts from Canada to Texas.

A second disease, again associated with high salinity and temperature, was recognized in oysters in the Delaware Bay in 1957. This organism was called "multinucleated sphere unknown," or MSX. Eventually MSX was found to be a sporozoan, a group of parasitic protozoa often found in insects, and was named *Michinia nelsoni*, although it was later renamed *Haplosporidium nelsoni*. How MSX gets around remains a mystery, as it does not appear to move from one oyster to another; it must have another, as yet unknown, host. The effect of these two diseases on the oyster fisheries of the North American East Coast has been devastating. Harvest in Chesapeake Bay alone dropped nearly 75 percent in fifteen years. Canadian Atlantic oyster fisheries have also suffered from disease, and MSX was diagnosed there in 2002.

As the importance of disease to oyster fisheries became recognized, the U.S. Congress instituted the Oyster Disease Research Program in 1991, bringing together scientists to study the diseases that led to the great losses to these fisheries. This collaboration resulted in rapid progress in understanding the causes of massive losses of oysters and to tests for the diseases, which made it possible to avoid the importation of diseased breeding stock. The scientists also launched a program to breed for disease-resistant oysters that has led to the hope that the impact of these diseases can be reduced in the future. Specially bred oysters are placed back into diseased areas, and their survival is monitored; those that resist disease then become breeding stock for the next generation.

On the Georges River south of Sydney, Australia, three generations of fishers made a good living from the oysters in these waters. In 1995 the oyster populations suddenly crashed, and within five years, the fisheries lost property, fishing vessels, and millions of dollars in income. Once again, a disease was involved. Eventually researchers learned that the disease had jumped from southern Queensland, where it was first seen in 1972, to New South Wales, about five hundred kilometers (three hundred miles) south, and wiped out the fishery. The disease was named "Queensland unknown," or QX. The culprit was found to be a protozoan, a member of a group called paramyxaprotozoa. QX suppresses the immune system of the oysters, making them susceptible to other diseases as well, and can lead to over 90 percent mortality in a single season. The spore invades the gut of the oyster and multiplies in its digestive gland, starving the animal to death within about forty days. Spores can survive for as long as three to four weeks in the sea, and (as with MSX) an unknown reservoir host is suspected. Australian scientists are also attempting to selectively breed the valuable Sydney rock oyster for resistance to QX disease.

As important as these shrimp, crabs, lobsters, crayfish, and oysters are to our diet, to the fishers who capture them and the farmers who raise them, and to the countries where they are sold, they are even more important to the marine and freshwater ecosystems where they have existed for millennia. The diseases that we have recently found in these invertebrates can be classified as emerging infectious diseases. They exhibit the same patterns as do frightening emerging diseases of humans such as severe acute respiratory syndrome (SARS) and bird flu: they are moved from one area to another by humans, are introduced into high-density populations, and lead to devastating mortality. Not only are these diseases of shellfish important to the humans that eat and farm them, but they may also have even greater effects on their native ecosystems if the grazing, filter feeding, and food-web functions of the shellfish are lost. We can only wonder what may happen then.

9

Deadly Hitchhikers

Crustaceans and Cholera

We often take the most important things in our lives for granted. When we drink a glass of water or flush the toilet, we would rather not think about what would happen if the water and sewage systems should suddenly fail. Yet only a small fraction of the world's population has these wonderful luxuries. The rest of the world suffers outbreaks of terrible diseases, including cholera. Recent research has led to the discovery that an innocent invertebrate may carry this disease and has also provided a simple solution to reduce its frequency.

Cholera has devastated humankind for many centuries. The term *cholera* or its equivalent appeared in ancient writings in Arabic, Sanskrit, Greek, and Latin at least 2,500 years ago to describe diseases causing diarrhea and vomiting. In Calcutta and Gujrat, India, there are shrines to the goddess of cholera. This disease was confined to India and especially Bengal (now Bangladesh) for most of its history. However, the goddess of cholera found a way of extracting a tax from the commerce that arose after invaders from the West began to attempt colonizing this part of the world. Europeans of the Age of Exploration arrived in India in 1498, and cholera appeared in Europe just five years later. Whenever armies marched into India, they underwent terrible attacks of disease that killed many soldiers. As British troops and commerce began to travel around the world during the late 1700s and early 1800s, so did cholera. When Muslim pilgrims traveled to Mecca, cholera killed many of them and went back home with the survivors.

In 1817 a terrible epidemic began in India, probably starting in Bengal. The epidemic moved throughout Asia by merchant ships and into Arabia with the British military, ending up as far north as Russia. This has been called the first cholera pandemic (*pandemic* is the term used for an epidemic that moves through many countries).

The first pandemic was followed by an even greater one in 1826. Again the disease moved out of India, spreading via ships and troops as far as Finland, Poland, Austria, and Germany. In 1831 cholera was introduced into London by warships, resulting in thousands of deaths. In 1849 it arrived in New York and moved across the United States. The invention of steamships that traveled rapidly between ports and the opening of the Suez Canal in 1869 increased the chances for cholera to move around the world. In all, six cholera pandemics were recorded from 1817 to 1923. Then the disease appeared to die down, becoming little more than a localized infection.

But cholera was only sleeping. In 1961 a new type, called El Tor, arose out of Indonesia, and in 1993 yet another new strain, called 0139, arose in India and Bangladesh. These led to the seventh and eighth pandemics. In 1998 almost 300,000 cases of cholera were recorded (with more than 10,000 deaths), but this estimate is probably low, as the most heavily infected countries do not report their cases. By 2001 the new 0139 strain of cholera had invaded fifty-eight countries. The goddess of cholera has awakened again.

During the fifth pandemic in the 1880s, a German physician and scientist named Robert Koch was studying the disease in Egypt and Calcutta, India. By a series of careful experiments, Koch proved in 1884 that cholera was caused by a bacterium that he named *Vibrio cholerae*. Thirty years earlier in England, contaminated drinking water had been shown to be the source of a cholera outbreak. Koch's observations clearly linked the disease to a bacterium in drinking water. In 1887 *Vibrio cholerae* was found in ballast water in

ships coming from France into New York harbor, and the ship was
forbidden to dump its ballast. This was the first time that laboratory bacteriology techniques were used to prevent an epidemic. England, Canada, the United States, and western European nations
began serious efforts to clean up drinking water and improve sanitation. The last cases in western Europe were seen in the 1920s—a
tribute to the incredible benefits of these improvements. But deaths
continue in Asia, the Middle East, South America, and Africa.

If you travel to a cholera-infected country in the warm months of
the year, and you are not careful about what you eat and drink, you
may contract cholera. You may have no symptoms at all, in which
case you can become a carrier and take the disease home with you.
But in the worst case, you will start to undergo watery diarrhea and
vomiting one to three days after infection. The diarrhea has a characteristic appearance called "rice water" because it resembles the
water in which rice is washed and cooked. Cholera victims can lose
up to twenty liters (approximately five gallons) of fluid in a single
day, leading to severe dehydration. Their eyes become shrunken,
they become very weak, and they suffer a terrible thirst. With no
treatment, they have only a fifty-fifty chance of survival. Treatment
is, however, remarkably simple—replace the fluids and wait. Simple
intravenous and oral fluid replacement kits have saved the lives of
many cholera patients. If antibiotics are available, they may eliminate the bacteria from the cholera victim's system.

Before asking what makes *Vibrio cholerae* so lethal, we must first
note that not all *Vibrio cholerae* strains are lethal. Many of these bacteria appear to live innocently in water without harming humans. The
strains that do cause disease are in two special groups, designated
01 and 0139. The lethal strains carry genes that code for toxins.

In 1884 Robert Koch had suggested that cholera produced a poison, but only in 1959 did two Indian scientists, S. N. De and N. K.
Dutta, produce diarrhea in rabbits using cell-free extracts of cholera
cultures, thus confirming that indeed something secreted by the bacteria was involved. Nearly one hundred years after Koch's suggestion that a poison is involved in cholera, scientists showed that puri-
fied toxin from *Vibrio cholerae* cultures produced rice-water diarrhea

in some truly dedicated human volunteers, confirming a toxin as the lethal agent. Eventually researchers found that the toxin is actually two molecules: a "B" molecule that binds to the cells and an "A" molecule that is responsible for entering the cells. These chemicals are secreted by the bacteria when they attach to the cells of the small intestine of the unfortunate victim. Through a complex interaction with the gut cells, the toxins induce the cells to secrete fluids and minerals. It is this gushing secretion of fluids that causes massive diarrhea, dehydration, and mineral deficiency and eventually leads to death if not treated. Cholera toxin is one of the most lethal substances known; a very tiny amount (a few thousandths of a gram) can lead to death.

Cholera outbreaks tend to occur in spring and late summer in Bangladesh. Scientists proposed that these outbreaks were somehow linked to warmer temperatures, perhaps attributable to sunlight or to chemical changes in the water. But an American marine scientist, Rita Colwell, realized that spring and late summer are also the times during which small floating organisms, called plankton, are at their highest populations in the water that the local citizens drink. This idea led to a passionate study of cholera by Colwell and her laboratory group that has extended for more than thirty years.

Colwell and her colleagues began to search for *Vibrio cholerae* on samples of organisms collected from water in Bangladesh. In 1973, they found the bacterium on small crustaceans (relatives of shrimp) floating in the surface water used for drinking. A flurry of activity followed. In 1983 Anwarul Huq and Colwell found *Vibrio cholerae* clinging to the mouth region and egg sacs of copepods, a tiny crustacean that was present in the millions in these waters. Each copepod carried up to 100,000 cholera bacteria. Later the group found the bacterium in the gut of blue crabs in brackish waters and estuaries off the North American coast, on the shed skins of aquatic invertebrates in Bangladesh, and attached to other forms of floating organisms such as algae. During the 1991–94 epidemic, when thousands of people were dying in Bangladesh, the researchers in Colwell's group learned that a single copepod plucked from Bangladesh waters carried enough bacteria to cause clinical signs of cholera.

The link between warm weather and cholera thus involved innocent crustaceans carrying a deadly hitchhiker, *Vibrio cholerae*. Crustaceans are reservoirs where cholera hides between epidemics, and when crustacean numbers grow during spring and summer, so does the pathogen. These animals are also the link that tied shipping to cholera outbreaks in the 1700s to late 1800s. Crustaceans with *Vibrio cholerae* were carried from one coast to another in bilge water, which was then discharged at the next port.

Why hadn't scientists found *Vibrio cholerae* on these organisms before? Culture methods, using agar plates and special bacteriological media, had been in use since the late 1800s. Eventually Colwell's group found that cholera has two forms: the one that multiplies rapidly in the human gut, leading to disease; and a second, nonculturable form, which clings to the skin of crustaceans. This form refuses to grow on laboratory media but instead is hiding, waiting to be swallowed by an unfortunate human. As methods became available to detect this nonculturable form, it was found in some surprising places, not just in areas with cholera outbreaks. For example, *Vibrio cholerae* was found in Australian rain-forest rivers, along the U.S. Gulf coast, and even in bottled water from a contaminated source.

Two Israeli scientists, Meir Broza and Malka Halpern, made another surprising discovery. Broza and Halpern were studying the chironomid midge, a nonbiting aquatic insect related to blackflies and mosquitoes that breeds in huge numbers in Israeli sewage treatment ponds. Female insects lay their eggs in glutinous masses on the edge of the ponds. Broza and Halpern found that eggs they brought back to the lab did not hatch. When they looked at these eggs under a microscope, they discovered a nontoxic strain of *Vibrio cholerae* attached to the egg masses. These bacteria had produced enzymes that degraded the gelatinous coating of the egg masses, destroyed the eggs and used the proteins to multiply. Further studies found both toxic and nontoxic cholera strains on chironomid eggs in India and Africa as well as in Israel. These insects, then, could be supporting cholera through their eggs. Even more disturbing is recent evidence that flying chironomid adults can carry *Vibrio cholerae* from place to place, helping the disease to spread.

The obvious technique for eliminating cholera, then, is to clean up the water. An attempt was made to do just that in Bangladesh in the 1960s by drilling wells for drinking water. However, many of these wells were found to contain arsenic, leading to the exposure of more than 30 million people to this toxic chemical. Bangladeshis returned to drinking surface water. Of course, if they could boil the water or treat it with chlorine, the problem would be solved—but firewood is a precious resource in this region, generally too expensive to use for boiling water, and chlorine treatment is not available. What, then, could be done to remove the cholera-carrying crustaceans from the drinking water? Colwell's group realized that the Bangladeshi women might have a solution very close at hand. They could filter crustaceans out of the water using the beautiful cloth that they wore every day—their saris.

Sari cloth is precious, and after it is worn and frayed to the point that it no longer can be used for clothing, it is kept for other purposes. One of these traditional uses is filtration of homemade drinks. Colwell's group persuaded women in three Bangladesh villages to fold an old sari cloth into eight layers and place it over the neck of their water pots. When they dipped water out of the river or pond and let the liquid drip through the sari cloth, the cloth filtered out any plankton in the water. After use, the cloth was rinsed in pond or river water and then in filtered water to remove the crustaceans, and finally it was left to dry in the sun. This last step, drying in the sun, was critical as it killed the bacteria. The villagers enthusiastically accepted this technique, and about 90 percent of them continued to use it for the three years of the study. The experiment led to a reduction in the number of reported cholera cases by about 50 percent, a remarkable achievement for such a simple solution.

Cholera is not gone. Crustaceans and probably insects will continue to carry it, and poverty, hurricanes, and floods such as the dreadful tsunami of December 2004 can lead to periodic epidemics. But until pure water becomes widely available to communities in developing nations, the simple sari cloth filter can continue to reduce the numbers of people sickened by this deadly hitchhiker.

10

Cloak and Dagger

Nematodes against Insect Pests

In their own way, Japanese beetles, *Popillia japonica*, are beautiful insects. Their metallic greenish brown bodies glint in the sun, their reddish wing covers and the white spots along their abdomens make them stand out as they cling to flowers and corn silk, strawberries and maple trees, apples and peaches, feeding away. Many mating pairs enjoy honeymoons in blossoms. But there are so many of them!

The Japanese beetle was accidentally introduced into New Jersey in 1916, and the beetles rapidly spread over thousands of miles. They ate nearly every useful plant in their path and were considered the single major pest of the time. By 1929 they had invaded much of the eastern United States and were rapidly making their way westward.

The adult beetles are ravenous feeders, but their young are, if anything, even more destructive. The adults lay their eggs just beneath the soil surface, and the white larvae, or grubs, burrow into the soil, seeking the roots of grasses and other plants. When enough grubs attack their roots, plants can undergo sudden death. Lawns and golf courses display very ugly dead spots. Trees have skeletonized leaves. Blemished fruit cannot be marketed. The Japanese beetle became a major target of insect control measures, and toxic arsenic-containing insecticides were sprayed over thousands of acres in an attempt to control this ravenous insect. But it continued its relentless march westward.

Rudolf Glaser, an entomologist at the Rockefeller Institute in New Jersey, was among many scientists looking for a solution to this enormous problem. One day in 1929, while digging up and examining Japanese beetle grubs, he found some that were dead or dying. When he opened them up, he was astounded to see that they were filled with very tiny, slender worms. Glaser had been trained not only in entomology, but also in parasitology and pathology, training that led him to immediately recognize these worms as nematodes (also known as roundworms). Glaser showed his nematodes to a scientist at the U.S. Department of Agriculture, Gothlob von Steiner, who realized that they represented a new species. Steiner named the nematode *Neoaplectana* (later changed to *Steinernema*) *glaseri* in Glaser's honor.

Glaser and his colleagues launched a campaign to use these nematodes to control the Japanese beetle. They started by attempting to culture the nematodes on an artificial medium, made from an extract of veal in agar covered with baker's yeast. After some modifications, this medium proved remarkably successful, and by 1937 Glaser was turning out as many as 140 million nematodes every day. The nematodes were first released onto a golf course in New Jersey, where they proved that they could infect beetle grubs. For ten years, between 1932 and 1942, Glaser and his colleagues successfully controlled beetle populations using this nematode. The project was making good progress, and nematodes were being released in many places, when, tragically, Glaser died in an accident in 1947 at the age of fifty-nine. Without Glaser's leadership, research on these nematodes ground to a halt.

A colleague of Glaser's, H. B. Girth, was able to recover the progeny of these worms from the same fields thirteen years after the nematodes were originally released, which is a remarkably long time for a biological control agent to persist. Although others searched for the nematodes in later years, they were not found again in these fields and appeared to be lost. However, in 1977, Wayne Brooks, a professor at North Carolina State University, discovered the same nematode in grubs of a different beetle in North Carolina, and research began again on this promising biological control agent. As recently as 2001, nematodes were rediscovered in Japanese beetle larvae dug up from turf in New Jersey.

Since most of us have never seen a nematode, we should first take a closer look at this group of organisms. Most nematodes are very small indeed. Many are microscopic, with a smooth, unsegmented cylindrical body and few external features to distinguish one from another. Nematodes look a little like curved sewing needles. Because they all look so much alike, only a few specialist scientists can identify nematode species, especially in the juvenile stages. Nematodes have digestive, reproductive, nervous, and excretory systems, but they are so primitive that they have no circulatory or respiratory systems. Nonetheless, nematodes are among the most successful and most numerous multicellular organisms on this planet. The soil is literally alive with them, so much so that an acre of cropland can contain as many as 80 billion nematodes. A classic story told to students is that if everything on earth but nematodes were dissolved away, the shape of the earth would still be visible because of the nematodes! They live everywhere from tundra to hot springs. Resistant to desiccation and other forms of stress, they can survive for long periods without food. They are responsible for recycling many essential soil materials. But many are also parasites, infecting hosts ranging from plants and insects to humans. A few nematodes (including one that infests grasshoppers) can grow up to several feet in length. Some (such as root-knot nematodes) cause diseases of plants, and a few (including a wilt disease in pine trees) are even carried into the plants by insects.

One tiny nematode, called *Caenorhabditis elegans* (*C. elegans* for short) has achieved great scientific fame as the experimental animal in which the phenomenon of programmed cell death, or apoptosis, was first observed. Apoptosis has been associated with diseases such as Alzheimer's, Parkinson's, and cancer. Thousands of scientists now work with this experimental animal. Every one of its cells and neurons has been carefully described, and it has become a model for testing the function and interactions of genes. In 1998 *C. elegans* became the first animal to have its genes fully sequenced. Three Nobel Prizes have been awarded to scientists working with this lowly nematode.

But perhaps the most familiar nematodes to us are those that take up residence in our bodies. Many children experience the intense anal itching associated with the very common and relatively innocuous pinworm. However, some nematode diseases are very serious or even fatal, and some of the worst of these are transmitted, or vectored, by insects. Pet owners will be familiar with dog heartworm, carried by mosquitoes. River blindness, elephantiasis, and filariasis—diseases that have ravaged entire regions of Africa—are caused by nematodes carried by blackflies and mosquitoes.

The association between insects, nematodes, and humans is a complex and long-standing one. Nematodes were first discovered in insects in 1602 by Ulisse Aldrovandus, who saw worms crawling out of a dead grasshopper. Since that time, we have learned that insect-infesting nematodes have three distinct stages: the egg; the juvenile stage which usually sheds its hard outer covering, or cuticle, four times as it grows; and the reproductive adult. As Glaser and those that followed him learned, the nematodes can go through all these stages for several generations within an infected insect until they use up all of its nutritional resources. At that point, the third-stage juveniles stop growing and do not shed their second-stage cuticle. This "cloak" protects them from the environment. This "dagger inside a cloak" is called a dauer (from the German word for "continuing" or "lasting") juvenile. The dauer then exits the remains of the insect and begins to hunt for a new host, using chemical cues in the environment that tell it where the insect may be located. The dauer can sense insect feces and even carbon dioxide given off by its prey, and it may lie in wait or actively search for a new host. When an unfortunate insect comes within range of the dauer juvenile, the nematode attacks through the mouth or anus and proceeds to drill directly into the body of the host. Other juveniles may follow the carbon dioxide trail into the spiracles, or breathing tubes, of the insect as well, entering the host in this way. In other species, the female nematode lays her

eggs in the environment, where they are eaten by an insect during its normal feeding. The offspring are already grown up to the second juvenile stage and are waiting inside the egg ready to attack the host. When the insect accidentally eats the egg, the juvenile hatches out of the eggshell and penetrates the gut, entering the body of the insect within a few hours. Then it completes its life cycle inside the insect.

Insects are not passive about being invaded by a foreign being, as Elie Metchnikoff learned early in his studies of immunology. Some nematodes can be encapsulated by blackened cells in an effort to shut them off from essential oxygen, the same melanization response observed by Louis Pasteur in injured silkworms. Insects can also produce enzymes to dissolve the parasite, and some insects may even actively resist the nematode by wiping them off their mouths or flailing wildly. These behaviors indicate that the insect is programmed to

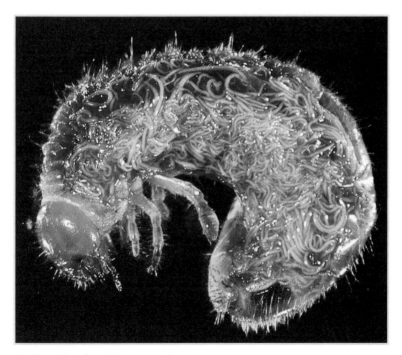

Beetle grub infested by nematode worms

respond to the presence of a dangerous parasite. Nonetheless, many nematodes are successful in setting up home in an insect host.

In 1955, Sam Dutky and his colleague W. S. Hough discovered a species of nematode that turned out to be an excellent agent for control of the codling moth, *Carpocapsa pomonella*, a serious pest of apples, pears, and several other fruits. The codling moth larva burrows into the fruit from the blossom end (opposite the stem) or directly through the side. The caterpillar tunnels out the seeds and core and leaves behind dark masses of fecal material, which ruins these fruits for market. At the height of their most destructive period in the 1930s and 1940s, the codling moth caused the abandonment of apple orchards in large sections of North America. Control depended on arsenic- or nicotine-containing chemical sprays, and these sprays had to be used very carefully to avoid contaminating the fruit. The nematode discovered by Dutky was in the taxonomic family Steinernematidae. Dutky called his nematode strain DD-136, and the species was named *Steinernema carpocapsae* in honor of the codling moth that it infected. DD-136 was soon found to be capable of infecting more than a hundred different insect species. It is a very active parasite; once it finds a host, it takes less than one hour to penetrate into the body cavity and begin its life cycle. During his dissertation research at Rutgers University in 1937, Dutky demonstrated that some nematodes carry a symbiotic bacterium, and he suggested that an antibiotic might be produced by the bacteria to suppress infection by other bacteria. This suggestion turned out to be prophetic of important future discoveries.

Although there are many different families of nematodes that infect insects at some stage of their life cycle, two of these groups have proven most interesting for possible biological control: the Steinernematidae, which includes *S. carpocapsae* and *S. glaseri*; and Heterorhabditidae, which are similar in having dauer juveniles but have a more complex life cycle. In 1976 George Poinar Jr., a professor at the University of California at Berkeley, found a new group of heterorhabditid nematodes parasitic in insects, and he called these nematodes *Heterorhabditis bacteriophora*. As the species name suggests, these nematodes carry a symbiotic bacterium with them into the insect. Poinar also found bacteria closely associated with *S. gla-*

seri, after this nematode was rediscovered by Brooks in 1977. In fact, bacteria had been seen in the intestines of nematodes infecting flies in Denmark forty years earlier, but their significance had not been realized at the time. As it became apparent that *Steinernema* and *Heterorhabditis* species had a great deal of potential against serious pests (such as the codling moth) that were difficult to control with chemicals, scientists began to focus on the complex life cycle of these nematodes and on ways to mass-produce them for sale as biological control agents.

Several insect-parasitic nematodes soon made their way into commercial products, with varying levels of success. *Steinernema glaseri* produced mixed results against the Japanese beetle, but *S. carpocapsae* proved to be a more promising candidate because it infected webworms, cutworms, armyworms, and wood-boring caterpillars, especially when these pests were in their soil-inhabiting stages. At least 90 percent of major insect pests spend part of their life history in the ground, either as larvae feeding on the roots of plants or as resting pupae, often during the winter. Soil-inhabiting insects are a difficult target for chemical insecticides, as the chemical must be applied to the soil and watered in to reach the depth where the insects are hiding. In contrast, the nematodes seek out their prey by cruising through the soil or lurking until an appropriate insect host comes into range. *Steinernema carpocapsae* stands on its tail in an upright position just under the soil surface. When a target insect passes by, it attaches to the host and begins its attack. Some species actively jump onto their prey, and not even rapidly moving insects can escape. Nematodes proved relatively easy to produce on artificial diets, in fermenters or on solid media. Some can be stored for up to five months at room temperature or a year in the refrigerator. All of these are attractive features for commercial production and field use. A different species, *S. feltiae*, became an important control agent for gnats infesting mushroom culture, while still other nematode species were found to work against vine weevils and mole crickets. Some imaginative products have been developed, including an attractant for cockroaches equipped with nematodes that attack the roaches when they come in to the bait.

In Australia, a nematode called *Deladenus* has been used to control a wood wasp of the genus *Sirex*. The interactions of these organisms form a complex web. The female *Sirex* lays her eggs beneath the bark of Monterey pine, an important forest product, using her long, needlelike ovipositor. At the same time, the female wasps also infect the tree with toxic mucus and spores of a fungus, and together these agents end up killing the tree. The wood wasp larvae eat both the wood and the fungus, and when they emerge as adults, they take the fungus along to the next tree. *Deladenus* nematodes do not kill the insect directly but rather invade the ovaries of the female wasp and render her sterile. When this wasp deposits her "eggs," instead of producing wasp larvae, the eggs produce nematodes! Each nematode larva then seeks out a wasp larva deposited nearby by a non-sterile female *Sirex* wasp, burrows in, and eats its host. But, cleverly, if no wasps are available, the nematode can also eat the fungus. Robin Bedding and Ray Akhurst, at the Australian Commonwealth Scientific and Industrial Research Organization, worked out mass-production techniques for *Deladenus* using a simple wheat-and-water medium. After the nematodes are ready to infect wasps, they are whipped into a gelatin foam that keeps them alive, and the foam is injected into holes in the trees. This inventive process results in nearly 100 percent infection of wasp larvae in the treated trees.

Under the right conditions, nematodes can be highly effective pest control agents, especially against soil-inhabiting or plant-boring insects, and they have the additional advantage over other biological control agents such as viruses in killing within twenty-four to forty-eight hours. Since these nematodes are specific parasites of insects, they pose no harm to plants, humans, birds, or other animals, unlike some of the chemicals that they may replace. Currently, more than 80 companies produce and sell insect-parasitic nematodes of five different species for use in greenhouses, home gardens, lawns, and agriculture.

As discussed earlier, symbiotic bacteria are associated with many of the nematodes. Curious scientists asked, how does the nematode live happily in the presence of teeming bacterial populations? Why is there only one species of bacteria present in the nematode-infected insect, when obviously there are many others in the surrounding soil and even in the gut of the host insect? Do the nematodes carry the bacteria from one insect to another?

The last of these questions was the easiest to answer. *Steinernema* dauer larvae enter the host through the gut or spiracles and penetrate into the hemocoel, the blood-filled body cavity. There they release bacteria of the genus *Xenorhabdus* through their anus. The bacteria multiply with extreme rapidity and kill the host within a day or two. The insect's body then becomes a factory for nematode and bacterial reproduction. The bacteria liquefy the insect body, feed on the juices, and multiply rapidly. The nematode then eats both the bacteria and the insect soup. When the nutrients are finally used up, dauer larvae are again produced. These larvae ingest bacteria and exit, seeking new hosts. If an adult female nematode becomes stranded in an exhausted host, she may donate her own body to her eggs, which hatch, feed on her body, and become dauer larvae that escape from their mother, carrying the bacteria with them. The nematode cannot complete its cycle without the bacteria, and the bacteria cannot enter the host without the nematode. The bacteria are therefore both symbiotic toward the nematode and pathogenic to the insect.

Heterorhabditid nematodes have a similar life cycle, but a different bacterium is associated with them. Instead of passing the bacteria through the anus, heterorhabditid nematodes regurgitate bacteria into their hosts. This bacterium, *Photorhabdus luminescens*, got its name because the infected host glows in the dark. The only other bacteria with this remarkable ability are found in dark aquatic situations such as the light organs of deep-sea fishes. The function of luminescence in *Photorhabdus* remains a mystery. Maybe the fluorescent molecule interferes with the production of a chemical called damaging reactive oxygen, or maybe it warns off predators or even attracts fresh insects as hosts for the next generation of nematodes and bacteria. Whatever the function of the glowing ability of *Photorhabdus*, it appears to be an essential and unique part of the interaction between the bacterium, the nematode, and the insect.

Scientists studying this interaction made many surprising discoveries. One interesting finding is that the most effective insect-parasitic nematodes are not surrounded by protective immune cells or by blackened, melanized cells, a response that would certainly occur if the insect was pricked by a needle. In addition, the bacteria are not inhibited by the complex of immune chemicals that insects generally produce in response to such a challenge.

These features caught the eye of scientists interested in immunity of insects, including Peter Goetz of Germany and Hans Boman of Sweden. Using Boman's cecropia moth pupa system for studying immunity, they attempted to identify the reasons why nematodes could bypass the insect's immune system. Goetz and Boman first produced *S. carpocapsae* nematodes that were sterile, lacking their symbiotic bacteria, and also cultured the bacteria separate from the nematode. When Goetz and Boman injected combinations of these organisms into cecropia pupae that had earlier been immunized against bacteria, they found a marked difference in the numbers of symbiotic bacteria required to kill the pupae, depending on whether the nematodes were present or not. If nematodes were absent, immunized cecropia pupae rapidly lysed the symbiotic bacteria. But nematodes killed the moths with or without bacteria; they just did it more efficiently with bacteria present. The scientists concluded that nematodes excrete an inhibitor that selectively destroys the immune proteins of the insect. Later studies by Randy Gaugler and others showed that at least one of these inhibitors is a protein carried on the surface of the nematode—yet another cloak produced by nematodes to ensure the success of the battle between nematodes, their symbiotic bacteria, and insects.

Other scientists focused on the symbiotic bacteria rather than the nematodes. Ray Akhurst, Noel Boemar, and others demonstrated that *Xenorhabdus* produces highly active antimicrobial substances that inhibit or directly kill competing bacteria that would attempt to take over the insect. Studies of *Photorhabdus*, the glowing bacteria associated with *Heterorhabditis* nematodes, have led to some extraordinary and potentially very useful discoveries. *Photorhabdus* has many faces during its life cycle. During its travels in the gut of the nematode, it is symbiotic, sitting quietly tucked into the wall of the nematode gut, waiting for its carrier to find an insect. But once the insect's body is penetrated by the nematode and the bacteria are released, they become active pathogens, killing the insect and converting it into nutritional material for both the bacteria and the nematode. *Photorhabdus* carries a complex series of genes that permit it to maintain these two very different roles. During its life cycle, *Photorhabdus* must do many things: act as a quiet symbiont, kill insects rapidly, multiply in insects, and repel other microorganisms that would take over the dead insect, especially normal gut bacteria. All the genes in *Photorhabdus* have been completely sequenced by a group of scientists at the Pasteur Institute. When the function of these genes was worked out, scientists learned that these bacteria make some unique gene products, including a series of toxins that kill the host within a day or two. Other proteins overcome the cellular and immune responses of the host, while still other gene products repel or kill competing bacteria, fungi, and even other nematodes that attempt to partake of the feast in the body of the dead insect. Then enzymes are produced to convert the body of the insect into food for both bacteria and the nematode.

Possible uses for the unique products produced by many of these genes, including new antibiotics, are currently being explored. Genetic improvement of the nematodes and bacteria is under way, in an attempt to increase the host range and efficiency of the nematodes. The intricate relationships between insect-pathogenic nematodes and their bacterial partners may lead to applications far beyond what Rudolf Glaser could have imagined when he discovered the dead and dying Japanese beetle larvae on a New Jersey golf course in 1929.

11

Bad News for the Good Guys

Diseases and Parasites of Bees

If you ask your friends to name the most important animal to humankind, they will probably answer that it is the cow, the pig, or maybe the chicken. However in terms of its importance to our food supply, there is one that exceeds the value of all these animals put together. That important creature is the bee. Because it is so important to us, we have learned a great deal about this insect over the centuries. Let's first explore the fascinating behavior of this important insect and then consider some serious threats to its existence.

Insects, especially bees, are critical to the reproduction of at least 85 percent of all plant species. Our food supply depends largely on the fruits and seeds from those plants (for example, apples and nuts). Either we eat those plants or plant parts directly, or we rely on products from animals that eat the plants (for example, cows eat clover hay). Flowers with petals produce pollen, the male component that fertilizes the female part of the flower (or pistil), leading to the production of seeds and fruit. Insects are the primary movers of pollen; without them, most seeds would not be produced. Plants go to great lengths to attract insect pollinators, producing brightly

120

colored flowers, attractive scents, and sweet nectar as a reward for
the insect's visit. They have evolved complex structures to deposit
pollen on visiting insects and to retrieve insect-borne pollen onto the
pistil.

The most important of these pollinators is the bee. You can distinguish a bee from its cousin, the wasp, by the presence of branched
hairs on its body and by its broad hind legs. These "tools" allow the
bee to collect and carry pollen back to its home to feed the next generation of bees. Bees apparently became distinct from wasps about
100 million years ago, at around the same time that flowering plants
appeared. Bees and flowering plants evolved together, with the result that many plants have highly sophisticated methods of attract-
ing bees, and bees have equally sophisticated methods of obtaining
pollen and nectar from plants.

Without pollination by bees and other insects, our diet would be
restricted to wind-pollinated plants, leaving us with meals consisting
of corn (maize), rice, wheat and other grains, and a few types of nuts.
We would have no apples, pears, oranges, cherries, plums, peaches,
strawberries, or raspberries, no pecans, beans, tomatoes, melons, or
squash, no broccoli, carrots, celery, or onions. Even plants that we
generally propagate by cuttings or rootstock, such as grapes and
potatoes, rely on pollination for the development of new varieties.
Our cows would have no alfalfa or clover hay, as these plants also
rely on bee pollination. When alfalfa and sweet clover were exported
from Europe into other continents without their pollinating bees,
farmers soon learned that the proper bees (in this case, bumblebees
or leaf-cutter bees) had to be imported as well. Indeed, beekeepers
often make far more income from renting their bees for pollination
of fruits and other crops than from the sale of honey. As a California almond grower remarked, "Without bees, my valuable almonds
would just be shade trees."

Among the thousands of species of pollinating bees, by far the
most valuable to humans is the European honeybee, *Apis mellifera*.
As a pollinator, this bee is a champion because it tends to return over
and over to the same type of plant, bringing back the right pollen
to fertilize the flowers. Among the many different species of honey-

bee, only *Apis mellifera* and its Asian relatives, *Apis cerana* and related species, have been domesticated.

Cave paintings dating back to 6000 BC show humans robbing bee colonies of honey. We do not know when bees were first domesticated, but drawings of beekeeping appeared on the walls of tombs in Egypt in 2400 BC. The bee appeared in the symbol for the city of Ephesus, accompanied by the Greek letters epsilon and phi, in the fourth century BC. Bees and beekeeping were mentioned in early writings by Aristotle (384–322 BC), Virgil (70–19 BC), and Pliny the Elder (AD 23–79). Honeybees are not native to North America but were imported by the early British settlers in 1622 to pollinate crops brought with them from Europe. The bees adapted well to North America and are primary pollinators for food crops, especially fruits, on this continent as well.

The honeybee that you see in your garden is almost guaranteed to be a worker, a nonreproducing female whose life is dedicated to taking nectar and pollen back to the hive. The relatively few male bees, called drones, have a short life with a single purpose — to mate with a queen — and are rarely seen outside the hive. Drones and workers alike start life as an egg deposited by their mother, the queen, in a cell of a wax comb deep within the hive. Only the queen produces fertile eggs. When the egg hatches, sister workers feed the larva a secretion called "brood food" produced from glands in their own bodies. Later they add pollen and honey to the diet. The pollen provides critical protein for the development of the worker larvae, while honey provides sugar for energy. The process of rearing larvae is remarkably efficient, with more than 90 percent of eggs generally developing into healthy workers.

At the end of about five days of development, the larva spins a silken cocoon within the cell, using salivary glands in her head. Her sister workers then cap the cell with wax. Inside this quiet cell, she changes into a pupa and about eleven days later emerges as a worker bee. From this moment to the end of her life, the worker bee is an essential part of the function of the hive and in some ways is a slave to its general well-being. Her first task is cleaning her own and other recently abandoned cells, and then she graduates to tend-ing growing larvae and cleaning and feeding the queen. Later, she will receive nectar from returning sisters, swallow it, and participate in converting it into honey. Enzymes added in bee stomachs convert the sugars in nectar into different sugars in honey once it is deposited within the wax comb cells. The worker may also help in building new wax combs and cells, act as a housekeeper by cleaning up debris and dead bees, and finally, move to the front of the hive and fan her wings hour after hour to ventilate the hive. She will also guard the hive from robber bees from other hives, wax-eating moths, and even animals as formidable as bears that seek to steal the honey. After two or three weeks of hive-bound activity, the worker may finally leave on her first foraging trip, seeking pollen and nectar. In this way, sister bees divide up the labor of the hive.

The queen, a little larger than her sister workers, is born in a very special, much larger cell, usually at the bottom or side of the comb. The workers feed her "royal jelly" (made by glands in their heads), which contains enzymes and hormones that permit the queen to reproduce. The queen spends about sixteen days in her comb cell before emerging. The workers become very excited when she emerges, but all may not go well for the new queen! If she is detected by her mother or a sister queen that emerged earlier, she may be killed on the spot. If the conditions are not right to start a new hive, she may be killed by the workers. But if she survives this first day, she has a far different life ahead of her than the workers. When she is ready to leave for her nuptial flight, a group of workers stands at the entrance to the hive and secretes a scent that will allow her to return safely home. Once she leaves the door, though, her sisters will not allow her to re-enter the hive unless she has mated. So off she flies to her first adventure!

Meanwhile, several days before the young queens emerge, the drones have been looking for desirable places to hang out, waiting for the ladies to arrive. Drones are produced just once a year for one purpose—to mate—and when mating is finished, the survivors are banished to die alone. Unlike queens and workers, drones are produced from unfertilized eggs. Because of this, the drones have only one copy of each chromosome and gene, while queens and workers have two copies. This situation, called haplodiploidy, is universal among bees, wasps, and ants. When a new queen finally appears, drones race after her, attempting to mate. During one or more mating trips, a queen will mate with half a dozen or more drones. The "lucky" drones sacrifice their lives to this task, literally exploding during ejaculation, an act that forces the sperm into the oviduct of the queen in a matter of seconds. Because the queen must be prepared to lay approximately one thousand eggs every day for as long as seven years, she requires much more sperm than a single drone can supply. She has the remarkable ability, shared by many insects, to store sperm until it is needed. Because of the queen's multiple matings, all of the thirty thousand or so workers in the hive have the same mother but not the same father. When her orgy of mating is over, the queen returns to her home hive, following the trail of scent provided by her worker sisters.

Now another adventure awaits the newly mated queen. For the past few days, workers have been scouting the neighborhood to find a home for a new hive. These scouts lead the way for a dramatic event—a swarming. Swarming is the only way in which honeybee colonies can multiply. Workers gorge themselves with honey, carrying supplies to feed the new colony until a hive can be established. Guided by the scouts, the queen takes off with a mass of honey-laden workers to establish a new colony. With luck, they will find a hollow tree or a wooden hive set up by a thoughtful beekeeper. These swarms are often sources of great panic for people who see the large numbers of bees buzzing in their trees or entering their attic, but generally the bees are not aggressive during this time—they are only interested in finding a new home. The bees immediately set to work building wax combs to receive eggs, pollen, and honey for the new hive.

Worker bees gather on the roof or side of the tree hole or in a human-constructed hive box and cluster together to warm the hive to the best temperature for working with the wax, about 35 degrees Celsius (around 95 degrees Fahrenheit). Workers remove flakes of wax extruded from their abdomens and then smooth it into very precise cells of exactly the right size (5.2–5.4 millimeters or two-tenths

of an inch) and hexagonal shape in the combs. The workers make a
central nurse area for the queen to lay her eggs, surrounded by cells
to be filled with honey and pollen, and leave just enough space between the combs for the workers to move about. Beeswax is remarkably strong: each pound (approximately half a kilogram) of wax can
support more than twenty times its weight in honey and represents
over 30,000 hours of bee work. Every pound of that honey required
about 65,000 trips to the hive and nearly 5 million visits to individual
flowers, thus representing flights that in distance are equal to several times around the world!

Until sugar made from sugarcane was introduced into Europe
in the early 1500s, the main source of sweetener was honey. The
British were engaged in beekeeping at the time of the Roman invasion, and around AD 400–500, special beehives made of cone-
shaped baskets called skeps were devised. However, harvesting the
honey from skeps without harming the bee colony was difficult, so

Langstroth moveable-frame beehive

in 1851 Reverend Lorenzo L. Langstroth designed the square white beehive that is familiar to most of us. The combs within the beehive are hung from wooden frames, which can be removed one by one. Swarming bees accept the hive readily, and it is very convenient for beekeepers to remove the honey for sale. The Langstroth hive is used worldwide wherever *Apis mellifera* is engaged in commercial honey production and pollination.

Harvesting honey and wax requires skill and a certain level of daring. Beekeepers cover themselves head to foot in a strong white cloth suit and wear a broad-brimmed hat covered with a mesh screen that permits them to see but keeps the bees away. Wearing heavy gloves, they carry a box with smoking embers, because centuries ago, humans learned that smoke had a calming effect on bees. When smoke is puffed into the hive entrance, workers are less likely to swarm and sting the beekeeper.

Honeybees have another talent that sets them above all other insects in their level of sophistication—they communicate with each other in several ways. One of these "languages" is chemical. If you are unfortunate enough to be stung by a honeybee, other bees will quickly surround you and attempt to sting you as well, especially if you are close to a hive. When the first bee stings you, she sacrifices her life to stop a perceived threat. Her barbed stinger sticks in your skin, pulling the poison sac from her body, and this sac continues to pulse, driving more and more venom into your arm. She also releases chemical signals, called pheromones, which shout to her sisters, "There is a dangerous intruder here!" These pheromones cause more bees to try to reach the spot and sting you again and again. Unless you are allergic to bee stings (which many people are), you will survive a few stings, but too many can be fatal. Your first response will be to run away, which is precisely what the bees want you to do! Carl Olson, an entomologist at the University of Arizona, described this situation well: "Beware of armed women. Everything that stings is female!"

But bees have an even more elegant method of communication. The first bee to venture from the hive on a spring morning finds a field of nectar-filled flowers. Within a short time, another will arrive, then several more, until the field is humming with workers collecting nectar and pollen. How do the workers back in the hive know that there is a valuable resource waiting for them in the field, and where that resource is? This question intrigued the German scientist Karl von Frisch, who spent many hours observing bees in glass-encased hives. Von Frisch eventually explained some of the clever ways by which bees talk to each other.

Von Frisch set dishes of honey-water out at various distances from the hive, and he marked the bees that arrived. Then he observed the behavior of the marked bees on their return to a glass-fronted hive. He discovered that the first bee to find the honey-water returns to the hive and dances in a vibrant circle on the comb, often turning and dancing in the opposite direction in another circle. Her sisters observe her closely, touching her with their antennae. Her dance tells them that there is a food source but not where it is. To tell the distance, the bee will begin tail wagging and dancing in a figure-eight movement. She will then make a straight run and continue the wagging, figure-eight dance. The sister workers deduce the distance to the source and quality of the nectar from the length of her straight run and from the speed and intensity of her dance. Since bees do not have a compass, they rely on the angle of the sun to tell direction. For instance, if the flowers are 30 degrees to the right of the sun, the bee will dance at a 30-degree angle to the right of vertical on the comb. The workers absorb this information, fly out of the hive, and accurately find the nectar source. This truly remarkable ability to communicate suggests a level of "intelligence" that is above that of most other insects. Our understanding of this communication is in large part attributable to the groundbreaking recognition by von Frisch and his colleagues that the movements of bees inside the hive are related to the activities of workers outside.

But as well organized as honeybee colonies are, they still suffer from diseases and parasites. As early as 700 BC, European beekeepers noticed that their hives suffered from disease. Aristotle, who was a beekeeper, described bee diseases in his *Historia animalium* in 330 BC and attributed these maladies to intoxication by bad nectar or pollen. Pliny the Elder, in the first century AD, described similar symptoms and also thought that they were the result of "sick" plants that the bees were visiting. The first description in modern times was that of Adam G. Schirack, who used the term *foul brood* in 1771 to describe a condition of larval bees, stating that it produced a "disagreeable stench." The bacterium causing foulbrood and experiments proving that it was contagious were described by Frank R. Cheshire and W. W. Cheyne in England in 1885. The connection between a bacterium and an insect disease was a novel idea at the time, as this was only two decades after Pasteur and others had conclusively shown the relationship between microorganisms and human diseases.

During the next decades, beekeepers in the United States gradually became aware that there were two different foulbroods. In 1904, a young American entomologist, Gershom Franklin White, began a study of foulbroods that would occupy his interest for more than thirty years. His publications contributed greatly to understanding bee maladies, as he used the best scientific techniques available at the time to tease out the differences between the two types of foulbrood that were appearing in bee colonies. He found that the first of these, which came to be known as American foulbrood, is caused by a rod-shaped bacterium called *Bacillus larvae* (now called *Paenibacillus larvae*). Despite being called American, the disease soon spread throughout the world, destroying a significant proportion of bee colonies. Several symptoms are characteristic of American foulbrood disease. Healthy honeybee larvae, deep in the hexagonal cells of the brood combs, are normally white. When infected with *Paenibacillus larvae*, however, they turn darker and darker brown until they melt in the bottom of the cell in a decaying mass. The foul smell of these diseased larvae is a distinctive odor similar to glue. When the larval cadaver eventually dries into a scale in the bottom of the cell, the mouthparts will often stick up, like a tongue. If the cell is capped over by attending bees, the caps will shrink. A clever, simple test was devised by beekeepers to detect the bacteria in dried cadavers. The scalelike larval remains are placed in diluted warm milk, and if *Paenibacillus larvae* is present, the milk will curdle in a few minutes because of an enzyme formed by the disease organism.

Paenibacillus larvae produces resistant resting stages called spores
that can remain alive and infectious for more than thirty years in
the scales of dead larvae and in the soil. These spores can be carried
from one hive to another by beekeepers. Bees tend to rob hives that
are not actively defended; in the process, they may also carry spores
back to their own hives. Spores can survive in honey, and a small
proportion can even survive heating to near boiling. The number
of spores required to infect and kill a larva is small, perhaps less
than one hundred. (By comparison, eight thousand to ten thousand
anthrax spores are estimated to be required to kill a human being,
so from the standpoint of the bee, American foulbrood is indeed a
terrible disease.) When spores are swallowed by a larva, they germinate within a few minutes into very active rod-shaped bacteria.
These rods take aim at the midgut cells of the larva and penetrate
these cells by taking advantage of one of the midgut's own mechanisms for uptaking nutrients, a process called phagocytosis. The
bacteria then move out of the gut cells into the body of the larva,
multiply at a rapid rate, and kill the larva within a few days.

Worker bees attending the larvae seem to be aware that a larva
is sick with American foulbrood disease. Indeed, some strains of
bees can detect the very early stages of infection, and workers of
these strains avoid cells that contain sick or dead larvae when stor-
ing honey, while other workers actively remove sick larvae and clean
the cells, lowering the spread of disease. Scientists in the late 1930s
and early 1940s observed that certain bee colonies were more resistant than others to American foulbrood. These resistant colonies
have two important characteristics, both of which seem to be inherited: the adult bees in resistant colonies are much more vigorous
at finding and removing the sick larvae, and the larvae appear to be
more resistant to infection. The brood food given to larvae by resistant bees also seems to contain inhibitors for bacteria.

Harry Laidlaw, an entomologist at the University of California
at Davis, developed techniques that permitted artificial insemina-
tion of queen bees. These techniques allowed beekeepers to select
the mates for their queens rather than to permit the random matings
that happen naturally. Using Laidlaw's techniques, resistant bees

were bred to lower the incidence of American foulbrood disease. A scientist at Ohio State University, Walter Rothenbuhler, became fascinated with the mechanisms that lead to resistance to American foulbrood disease. From 1951 until the mid 1970s, Rothenbuhler and his students found some inherited behaviors that lead to susceptibility or resistance, including the propensity to seek out and destroy sick larvae. They also found physiological differences among bee strains and explored the mechanisms by which the bac-

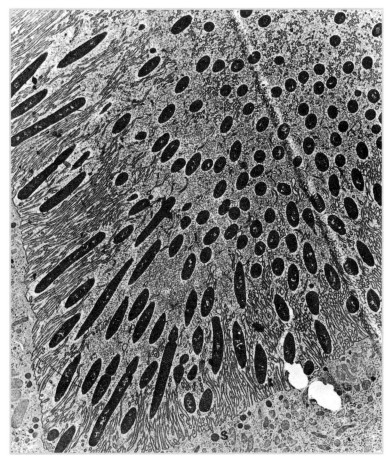

Paenibacillus larvae penetrating midgut cells of a bee larva
(electron micrograph)

terium invades the bee. Despite the efforts of these and many other
scientists and beekeepers, this disease continues to be a very real threat to the pollinating and honey production activities of *Apis melli-fera*, requiring constant vigilance to ensure that the disease is not spread by either beekeepers or bees. In the United States, American foulbrood–diseased hives are still generally burned to kill the bacteria.

In 1771 Adam Schirack, an excellent observer, determined that there were two different maladies affecting his bees. We now know that the second disease, called European foulbrood, is (like American foulbrood) caused by a bacterium. Early scientists thought that European foulbrood was caused by rod-shaped bacteria called *Bacillus alvei* or *Bacillus pluton*, but eventually Lucian Bailey and other scientists found that it was caused instead by a different organism, called *Streptococcus* (now called *Melissococcus*) *pluton*. A unique characteristic of European foulbrood is that sick larvae die coiled up in their brood cells. This disease is found on all continents where bees are cultured, but it is not considered as dangerous as the American form.

A simple test is used to distinguish between the two foulbroods. A matchstick or toothpick is dipped into the bottom of a cell containing a dead larva and pulled slowly out. If a long, unbroken thread can be drawn out, the disease is probably American foulbrood; if the thread breaks short and the larval remains are granular, it is probably European foulbrood. A European foulbrood–infected colony may also have a distinctive sour odor. Unlike *Paenibacillus larvae*, *Melissococcus pluton* lives only within the larval gut, where it grows rapidly, fills the gut, and greatly weakens or kills the larva. Nurse bees recognize the sick larvae and remove them but also spread the disease in the process. As with American foulbrood, some colonies appear to be more resistant than others. Antibiotics and ethylene oxide have been used to control the disease. If it is not too widespread, bees can overcome a mild case of European foulbrood by removing the sick and dying larvae.

Bees are also susceptible to an array of other pathogenic organisms, including protozoa, fungi, viruses, and bacteria, as well as

larger parasites. The year-round life of the bee colony, the large number of individuals within a colony, the warm temperatures maintained in the hive, contact with the outside world during foraging,
and the tendency of bees to rob honey from other hives all contribute to infection. The movement of bees around the world by humans
has also inadvertently led to the spread of bee diseases.

One of these pathogens is *Nosema apis*, formerly thought to be
protozoa but currently believed to be a member of a phylum called
Microspora within the kingdom Protoctista. Microspora are largely
pathogens of insects, producing resistant spore stages that permit
them to survive in the environment until they are eaten by a susceptible insect. Once in the gut of the bee, the *Nosema apis* spore fires out
a coiled fiber, called a polar filament, into a midgut cell. An infectious unit runs down the hollow filament into the cell and infects it.
These pathogens grow only inside the cells of the midgut and do not
kill the bee but shorten her life instead. This often leads to a lowering of the number of active bees in the colony but only rarely to the
death of the colony.

Bees also suffer from nearly a dozen viruses, including one called
sacbrood. This disease was recognized by White in 1917, well before the invention of the electron microscope and other sophisticated
equipment for detecting viruses. As the name implies, after diseased
larvae die, their skin turns into a tough sac filled with fluid containing a large number of virus particles. Bees are also infected by fungi,
including *Ascosphaera apis*, which causes a disease called chalkbrood.
When spores of this fungus are eaten by the larvae, they germinate
in the gut and penetrate into the body. After the larva dies, spore
masses form on the outside of the dead larva, leading to the chalky
appearance. Occasionally, even nematode worms have been found
parasitizing bees. All told, more than thirty different infectious diseases are known to afflict the hard-working honeybee.

As if honeybees did not have enough problems with diseases,
they also have to deal with several parasites, including mites, which
are close relatives of ticks. Two different types of mites were first
seen in honeybee colonies at the beginning of the twentieth century.
The first, *Varroa jacobsoni* (now called *Varroa destructor*, an appropri-

ate species name), first appeared in Asia in 1904. It moved into West-ern Europe via importation of bees from eastern Russia around 1970 and was first described as a pest there in 1975. *Varroa* is a round, flat, eight-legged arthropod that resembles a tiny crab. It hops onto the outside of a worker bee and moves quickly to hide in a fold of the bee's body. Its special wedge-shaped head-plus-thorax can slide under the tough outer layer of the body segments and into the soft cuticle underneath, where it inserts a very sharp pair of mouthparts with tiny hooks.

Apis cerana, the Asian honeybee, is much more resistant to *Varroa* than is the European bee, *Apis mellifera*. Ying-Shen Peng, study-ing bees in Beijing, China, described the remarkable mite-avoidance behavior of the Asian bee. Asian worker bees are very alert to the presence of this parasite, and if a bee is attacked by a mite, she im-mediately twists around and attempts to brush the mite off with her legs. If the mite manages to get into a groove in her body, the bee begins to dance and vibrate her abdomen, which leads her sister workers to run to her and examine and groom her carefully. The in-fested worker will pull up her abdomen and open her wings, as if to say, "It's right there—get it off!" The bees then bite and chew the mite and are very successful in removing nearly all of the mites. If the bee does not rid herself of the mite, as often happens with Euro-pean bees, she will have a shortened life span and may actually die from secondary infections, bacteria, viruses, or fungi that take ad-vantage of her weakened condition.

Female mites use the hormones of the host bee to induce their own egg production. Mites then leave the worker bee and seek a cell with a bee larva nearly ready to pupate. The mites hide in the brood food at the bottom of the cell until the worker bees cap it with wax, whereupon they begin to feed and lay eggs on the larva. When the bee larva finally emerges as a worker, the family of mites will travel out with her and infest other bees in the hive. If the beekeeper does not clean the colony of infested bees and use some form of chemical control of the mites, a heavily infested colony will die within a few years.

Varroa is only one of many mites that infest bees, some more damaging than others. A very tiny mite, *Acarapis woodi*, infests the tracheal system of the bee. Insects do not have lungs or oxygen-carrying red blood cells; instead, they move air through their bodies using a series of pipes called tracheae. These tracheae open to the outside along the sides of the insect through tiny openings called spiracles. Tracheae branch into smaller and smaller tubes until they carry oxygen to all the organs in the insect's body. The tracheal mite feeds and reproduces in these tubes.

Acarapis woodi (formerly called *Tarsonemus woodi*) was first described by J. Rennie in 1921. Beekeepers on the Isle of Wight off the coast of England saw bees that appeared to be partially paralyzed crawling around their hives, and many colonies died soon afterward. This tiny parasite sneaks into the spiracles and infests the larger trunks of the tracheal tubes. Each invading female lays many eggs, which rapidly hatch into more mites that infest the same tube. The mites feed by piercing the wall of the trachea and feeding on the blood of the bee. Tracheae can be crowded with mites, obviously causing distress to the bees and potentially lowering their resistance to other diseases. This pest was first seen in North America in Guadalajara, Mexico, in 1982, and it has since moved through most of the United States. In conjunction with other diseases, mites have caused major losses of bee colonies.

It can be argued that importation of the honeybee into North America was a good thing, permitting the pollination of many crops. However, the original imported bee, a dark subspecies from northern Europe, was aggressive and hard for beekeepers to control, and it was susceptible to several diseases. In 1859 a more gentle bee, the golden Italian form, was introduced and became the preferred type for American beekeepers. This bee was, however, also quite susceptible to diseases and parasites, leading to the search for a pest-resistant strain. In an attempt to solve this problem, a Brazilian entomologist, Warwick Kerr, imported a strain from Africa in 1957. His goal was noble: to crossbreed the African bees, which were reportedly more resistant to disease and parasites, with European bees. Kerr's experiment turned out to have tragic results.

The African bees almost immediately escaped from their enclosures and established colonies in the wild, taking up residence in tree holes, nooks in rock walls, animal burrows, and houses. They moved rapidly through Brazil and northward into Venezuela, Colombia, and Mexico, and they entered the United States in Texas in 1990. They have now colonized much of the southern United States.

The African bees soon became known as "killer bees," because they rapidly attack any person or animal that they perceive as a threat, swarming onto the enemy and stinging en masse. Many humans and domestic animals have been killed by these bees. African bees will take over and evict or become incorporated into a gentler bee colony, a behavior that poses a real threat to the beekeeping industry and its associated pollination activities. The beekeeper can return to a hive one day and find a very different population. Thus, beekeepers are forced to check their queens every few weeks to be certain that an invading African queen has not evicted the gentler queen. African bees are able to move without the assistance of humans, to live in the wild, to multiply faster and resist diseases and parasites better than European bees, and to take over beehives. For these reasons, the African bee unfortunately appears to be here to stay. In the words of Pliny the Elder from two thousand years ago, "Out of Africa, always something new."

12

The Mysterious Reappearing Fungus

Gypsy Moth Pathogen

It seemed like a good idea at the time.

In 1868 Leopold Trouvelot, an astronomer at the Harvard Observatory, was pursuing his hobby: the production of silk. As noted in earlier chapters, silk was an important agricultural product in Trouvelot's home country, France, in the mid 1800s. Trouvelot was aware that a serious disease, the black spot (or pebrine) disease of the silkworm, was threatening to destroy the French silk industry. Louis Pasteur had begun his work at Alés in the south of France three years earlier and would publish his groundbreaking book on these diseases in 1870. Trouvelot reasoned that some species other than the domestic silk moth, *Bombyx mori*, could potentially be used to produce silk. Or perhaps native moths could be crossbred with other species to produce disease-resistant silkworms. To test this possibility, Trouvelot imported egg masses of a silk-producing caterpillar from Europe to use in his experiments.

Trouvelot's imported caterpillars were called gypsy moths, *Lymantria* (formerly *Porthetria*) *dispar*. In Europe, the gypsy moth was known to be destructive to forest trees but was not considered a major pest. Around 1869, a few of these insects escaped from Trouve-

lot's home at 27 Myrtle Street in Medford, Massachusetts, into the
surrounding forest. As a good scientist should, Trouvelot notified
other scientists about this accidental release. He received no response from local officials about the release of an insect, and the
moths seemed to have disappeared. Trouvelot then gave up his entomological experiments and concentrated on astronomy, returning to
France in 1882.

About twelve years after the gypsy moths had escaped, caterpillars suddenly appeared in massive numbers on a single street in
Medford, near Trouvelot's home, where they seemed to be confined
for another ten years. Although the local citizens were upset by these
pests, little was done aside from cleaning up their messy bodies and
burning them.

Then in 1889 a massive plague of caterpillars practically over-
whelmed the small town of Medford. Trees were stripped of leaves,
while house walls and tree trunks, fences, and sidewalks were covered with crawling caterpillars, some of which invaded houses and
ended up in beds and food on pantry shelves. Citizens walking out-
doors were assailed by caterpillars dropping down on silken threads
from overhanging trees. These caterpillars would aggressively climb
up any vertical object: a telephone pole, a fencepost — or even a person's leg! Shade and fruit trees were stripped bare of leaves and
killed. Woe be to the housewife who hung out her laundry to dry;
her clean sheets would soon be covered with caterpillars and stained
with excrement. The caterpillars invaded about 350 square miles of
Massachusetts, and this time they caught the attention of entomologists and government agencies.

Specimens were sent to an entomologist at the Agricultural Experiment Station in Amherst, Massachusetts. There Maria Fernald,
an avid collector of moths, recognized that this was not a native
caterpillar but was the "famous gypsy moth of France." The citizens
of Medford immediately held a town meeting and began a concentrated attack on the insect. Because this was an exotic pest and posed
a real threat, the State of Massachusetts set out to eradicate the
gypsy moth in 1890. Egg masses were scraped off the trees, leaves
on the ground where egg masses might be found were burned, and

burlap sacks were placed on the trunks of trees to induce caterpillars to hide under them. These sacks were then removed and burned.

Modern chemical insecticides had not yet been developed, and many of the pesticides used at the time, including creosote and Paris green, were frightfully toxic and would not be permitted today. Paris green contains copper and arsenic as its active ingredients, and it was used in such high quantities that the leaves of the trees turned brown. Understandably, citizens of Medford objected to the heavy use of this toxic insecticide and even threatened the workers running the sprayers. Their concern was not as much about the toxicity of the chemical, however, as for damage to their trees, for when Paris green was replaced in 1893 by lead arsenate (which also contains two highly toxic chemicals), they did not oppose the application of this chemical compound, as it did not damage their trees. Spray equipment was rapidly developed that used high-pressure water to reach the tops of trees. The chemicals were applied using solid streams of poisoned water thrown up with great force. Water was suctioned from mountain streams to mix with the lead arsenate, which of course eventually ran back into those streams. The investment in this project was great; Massachusetts set aside $25,000 each year, the equivalent of about $1.2 million today. By 1901 extermination appeared to be close, the gypsy moth was no longer seen as a problem, the public was no longer crying for assistance, and so the legislature stopped support of the extermination of the gypsy moth. In hindsight, what a mistake that was!

It is easy to see why Trouvelot thought this might be a good silkworm. The gypsy moth is a member of the family Lymantridae, known as tussock moths because the larva bears tufts of hairs on its back. The adult female gypsy moth is rather attractive, a light buff color with some dark zigzag markings on her wings, which have scalloped edges. Like the true silkworm, she cannot fly. The male is smaller and darker brown, with feathery antennae. He flies up and down tree trunks seeking a receptive female, who signals her presence with a series of chemical perfumes called sex pheromones. The female waits on the tree trunk for a male to approach, and the two mate. A few days later, she lays approximately four hundred eggs in each of several egg masses on the tree trunk or the dry leaves below. She then dies, only a week after mating. The egg masses are covered with hair from the female, which protects them from many predators. Although the eggs are laid in midsummer, they do not hatch until the following spring, producing only one generation per year. Young caterpillars feed on leafy trees but as they grow, they can voraciously attack evergreens as well. Finally, the fully grown caterpillar spins a silken cocoon on tree trunks or under bark and changes into a pupa. In midsummer it finally emerges as a moth.

The gypsy moth is referred to as an episodic insect because it normally undergoes boom-and-bust cycles, depending on the weather, wind, temperature, and available food. If larvae are at high density, they will extend their life span by adding molts. But the survivors of such a high-density situation have given themselves an advantage; they are larger and can lay more eggs as adults. Therefore, crowding actually may lead to a higher reproduction rate in the next generation. Larvae that have freshly hatched from eggs are tiny and are often not noticed by the public. They climb to the tops of trees, spin a thread, and disperse. They then proceed to eat and grow until they are nearly a thousand times as large as when they first hatched. In the process, each caterpillar can eat a full square meter (slightly more than a square yard) of leaves.

The gypsy moth is a classic invasive species. It arrived in Massachusetts without its native parasites, predators, and pathogens. The trees that provide meals for the larvae in Europe and Japan have, over the millennia, developed mechanisms to resist this insect. In North America, however, the moth found more than three hundred species of related trees and shrubs that have no such defenses. If a tree is totally defoliated (that is, all its leaves are eaten) once or twice, it can usually survive. However, by the third time it is defoliated, its nutrient reserves will have been exhausted, making it highly vulnerable to other boring insects and plant diseases and often resulting in its death.

Complete eradication of an imported pest (or for that matter a native pest) has rarely been accomplished. However, at the time of its initial outbreaks, the gypsy moth was a good candidate for eradication. It was a definite nuisance, so citizens affected by the pest would make sure that it would stay on the minds of local and state officials. The female moths cannot fly, so the species spreads mainly by caterpillar movement, which is fairly slow (although not so slow as might be expected). The caterpillars can be readily identified and do not hide, and the early outbreaks were in a well-populated area where they could be easily observed. But when the gypsy moth numbers became low enough that the public no longer complained, the Massachusetts legislature felt that the funds could be better spent on other projects. If this decision had been delayed for even a few years, our story might have ended here, and eradication might have been successful.

Over the next few years, the gypsy moth took advantage of the situation and rapidly expanded its range. By 1905, it had invaded more than two thousand square miles in Massachusetts and then proceeded to march into New Hampshire, Connecticut, and Vermont. Because the female moth cannot fly, movement from one place to another was apparently the result of humans moving timber or even automobiles that had eggs glued to them. Small caterpillars are covered in hollow hairs, making them buoyant, and they spin out a long silken thread that permits them to be blown by the wind, sometimes for several miles. These were the major ways by which the insect moved far beyond Medford.

Once the gypsy moth had moved out of Massachusetts into neighboring states, it became an insect of interest to federal as well as state agencies, and the U.S. Department of Agriculture began an attempt at biological control. Parasitic wasps and flies were imported from Europe and Japan, reared in the laboratory, and released into areas infested with gypsy moths. During this project, from 1905 to 1933, more than forty such natural enemies were released, and nine of them became established. Most attack only gypsy moths, and there is little evidence that they have harmed other insects such as butterflies. In 1920 a second group of gypsy moths were accidentally introduced (on pine trees from the Netherlands) into New Jersey, but these were successful eradicated by 1931. Nonetheless, the gypsy moths marched on through the eastern United States. From

1906 to 1920, they traveled about six miles per year, moving into New York and Ohio.

In 1923 a barrier zone was established along the Hudson River and Lake Champlain, from close to New York City to the Canadian border, a zone 250 miles long and 25–30 miles wide. Scouts were sent out to search for caterpillars in this zone, with instructions to destroy any that they found. The barrier appeared to work until the 1960s, although small infestations had already been found outside the barrier in New York, Pennsylvania, and New Jersey. The barrier probably slowed the spread of the gypsy moth as long as it was maintained.

Immediately after World War II, two new weapons against insects appeared on the scene: DDT and the airplane. DDT was first synthesized by a German scientist in the late 1800s, but not until 1942 did its insecticidal properties become known to American scientists. It was a miracle in its day, as its long residual life and high toxicity to insects led to its use in controlling malaria-carrying mosquitoes, typhus-carrying lice, and many important agricultural pests. Efficient use of airplanes for spraying, also developed during the war, meant that the gypsy moth could be attacked over hundreds of acres of remote forest in a single day. In 1949 all of Cape Cod was sprayed, in 1951 all of Nantucket Island, and in 1956 all of Martha's Vineyard, with the goal of totally eradicating the gypsy moth. In 1957, 3 million acres (1.2 million hectares) were sprayed with DDT. Eradication was once again discussed as a possibility. As a result, other forms of control, including biological control, dropped from the list of options until the negative effects of DDT became known in the late 1950s. The publication of Rachel Carson's *Silent Spring* in 1962 led to public recognition of the hazards of widespread aerial spraying of DDT for gypsy moth control. Public protests led to the replacement of DDT with other chemicals, although homeowners were still very concerned about aerial spraying of their backyards with insecticides. Without control of this insect on city property, however, widespread eradication was impossible, and so the gypsy moth persisted and continued to move westward. In 1971, the largest area of defoliation ever seen was recorded: nearly 20 million square miles (50 million square kilometers). By 1981 the insect had been found in the midwestern states of Wisconsin and Michigan and even as far west as the Pacific coastal state of Washington.

As discussed in previous chapters, a critical step in controlling an insect is the ability to rear it in the laboratory. The gypsy moth remained resistant to such rearing until artificial diets and rearing techniques were perfected in 1966. Several pathogens — including a nuclear polyhedrosis virus and a special strain of *Bacillus thuringiensis* — that showed good activity against gypsy moth were then developed. The virus was found naturally occurring in the gypsy moth, and it led to cycles of population crashes, described first by Charles Doane and his colleagues at the Connecticut Agricultural Experiment Station in the 1970s. The virus was eventually developed into a registered insecticide named Gypcheck, but because of the expense of producing virus in living caterpillars, this insecticide was never widely used. The *Bacillus thuringiensis* product was widely used, but concern for the possible death of rare butterflies and moths has led to reduced use of this microbial control agent as well. Other clever techniques — including an artificial insect sex hormone named Disparlure to attract and confuse males, and the release of sterile males to mate with females and produce infertile eggs — were also developed. Slow progress was made, but the search continued for a safe, effective control of this devastating pest.

Soon after the invasion of the gypsy moth began, entomologists realized that if they wished to find a good biological control agent they should go to the source (namely, to Europe and Asia, where the pest had originated). In 1908, gypsy moth larvae in Massachusetts were found to be dying from a fungus disease, suggesting that such a disease might be a useful biological control agent. Funds were provided by an anonymous donor for George P. Clinton, a plant pathologist from the Connecticut Agricultural Experiment Station, to travel to Japan the next year to collect diseased gypsy moth larvae. Clinton brought back only two fungus-infected larvae, but he succeeded in infecting locally captured larvae in the laboratory. In 1910 and 1911, these infected caterpillars were released near Boston, Massachusetts. Unfortunately for the fungus-importation attempt, in 1911 a massive virus outbreak occurred in the caterpillars, and the fungus was not seen again.

Japanese scientists began to find outbreaks of fungus disease in gypsy moths from 1954 until the mid 1980s. They recognized that this disease closely resembled *Entomophaga aulica*, a fungus that kills the brown-tailed moth, which is in the same family (Lymantridae) as the gypsy moth. Similar disease outbreaks were reported in Korea, China, Russia, and occasionally Europe. In 1981 Richard Soper and colleagues from Cornell University returned to Japan and brought back more fungus-infected caterpillars. This time the Cornell group succeeded in culturing the fungus on an artificial medium consisting of egg yolk and other ingredients on agar plates. They named it *Entomophaga* ("insect eating") *maimaiga* (the Japanese name for gypsy moth) but recognized that it was obviously quite closely related to *Entomophaga aulica*. Another Cornell colleague, Ann Hajek, released the fungus in western New York and northern Virginia in 1985 and 1986—again with no success. The gypsy moth appeared to be winning the war.

During an unusually humid spring in early June 1989, a massive die-off of gypsy moth larvae suddenly occurred in Connecticut. By July, caterpillars were dying in five other eastern states as well. Hajek and her colleagues were able to culture the same fungus, *Entomophaga maimaiga*, from these dying and dead caterpillars. Somehow this fungus had taken a foothold in the United States but, oddly, not at any of the locations where it had been released in 1985 and 1986! Hajek felt that the disease outbreaks were not likely to have been related to her intentional releases, as the first appearances were far from her release sites in New York and Virginia. A year later, the disease was found in Maine, Delaware, Maryland, and Ontario, Canada. The outbreaks of disease in 1989 and 1990 appeared to be linked to heavy rainfall, leading to the high humidity that favors fungal growth. But not all areas with gypsy moth populations had fungus die-offs, and often die-offs would occur even when rainfall was low.

The actual source of the lethal fungus remains a mystery, but there is one interesting hypothesis, put forth by Ronald Weseloh of

the Connecticut Agricultural Experiment Station. He and one of his
colleagues made a trip to China to search for natural enemies of the
gypsy moth in 1982. Although they did not bring fungus-infected
caterpillars back to the United States, the first outbreak of disease in
1989 occurred in Connecticut close to the homes of these scientists.
Weseloh proposes that he or his colleague may have transported the
fungus in mud on their shoes. Hajek has also suggested that the virulent strain of *Entomophaga maimaiga* may indeed have been independently, and perhaps accidentally, imported into the United States
from Asia. Genetic studies of the fungus showed that all the strains,
regardless of where they were found, are very closely related. This
information suggests that all the strains originated from one or a few
introductions.

The ability to culture the fungus on an artificial medium in the
laboratory allowed the Cornell group to determine that this fungus
has many faces. Using medium designed for insect cell culture as
well as commercially available agar-based media for bacteria and
fungi, Soper and his colleagues were able to produce several different forms of the fungus. After the infected larva dies, *Entomophaga
maimaiga* pushes rootlike conidiophores out through the cuticle of
the larva and releases pear-shaped spores called conidia from the
surface. These infective conidia are actively discharged from the
original host and may land on the cuticle of a nearby larva. They
may also be carried in the wind to caterpillars on another tree. There
they stick fast to the surface of the insect and begin to penetrate
the cuticle using both mechanical force and enzymes to dissolve
away the protective layers of the cuticle. Once inside, they trans-
form into multinucleate protoplasts, which have no cell wall and
look rather like protozoa. These protoplasts, multiplying rapidly,
migrate through the body of the host. For several days, the caterpillar appears to be unaffected. Then the larva slows its feeding as its
bodily nutrients are used up by the fungus, until it finally stops eating altogether. About two days later, it suddenly becomes paralyzed,
and about an hour after that, it dies. During the next day, the cycle
will be repeated, with the formation of new conidia to infect the next
generation of caterpillars. Thus the infection of only a few caterpil-

lars early in the spring gradually leads to more and more infectious spores being shed until ideally a massive die-off of the gypsy moth larvae occurs.

However, *Entomophaga maimaiga* has yet another weapon in its arsenal. In June or early July, some infected larvae produce a different form of spore, called an azygospore or resting spore. Resting spores are larger than conidia, are round, and have very thick double-layered cell walls and two nuclei. These spores remain within the larval cadaver and are highly resistant to cold, heat, and drying. This cadaverous time bomb becomes the source of infection for caterpillars in the next season or perhaps even many years later when the cadaver disintegrates and the spores enter the soil. The resting spores refuse to germinate immediately even if placed in a warm incubator. After several months at freezing temperatures, however, perhaps in the following March after a rain shower, each

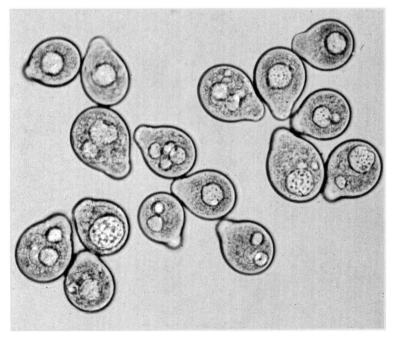

Entomophaga maimaiga spores

resting spore will produce a single infectious conidium. This spore must find a susceptible caterpillar to start the cycle again. These resting spores are what may have been carried to Connecticut and to other parts of the United States on shoes or equipment, eventually spreading the fungus across the range of the gypsy moth.

An important question always arises during the proposed introduction of a biological control agent: what other insects might be affected by the pathogen, predator, or parasite? Extensive studies of *Entomophaga maimaiga* showed that it affects only the gypsy moth and a closely related insect pest, the Douglas-fir tussock moth. Very few larvae of other moth species have been found to be infected by this fungus during outbreaks in the field. Because it is apparently highly specific to the gypsy moth and closely related pests, the fungus is a truly beneficial biological control agent. The high level of

Gypsy moth caterpillar killed by *Entomophaga*

susceptibility of the gypsy moth to this fungus also suggests that the insect had not had contact with the disease over many years, so the fungus probably had not been associated strongly with the caterpillar before the insect was introduced into the United States.

This fungal pathogen has been released in many areas of the United States that are afflicted by gypsy moth populations. Caterpillars containing resting spores are used to inoculate each new area. These experiments resulted in establishment of the fungus at many sites, even in several areas where it had not been released. Once again the transportation of the fungus on shoes and equipment was suggested to be the cause of this unexpected result.

Although further research on this promising pathogen may lead to even better control, we must remember not to repeat the mistake of the Massachusetts legislature in 1901 and give up prematurely. Ronald Weseloh stated the issue this way: "The history of gypsy moth control has been littered with missteps, accidents, good intentions turned wrong, and, fortunately, serendipity. I believe the gypsy moth case is a good example of how careful consideration of the consequences of actions would have made big differences in the outcome."

13

The Sixth Plague of Egypt

Fungus and the Desert Locust

The sixth plague was brought down upon Egypt to persuade the pharaoh to free the Hebrews:

> And Moses stretched forth his rod over the land of Egypt, and the Lord brought an east wind upon the land all that day and all that night, and when it was morning the east wind brought the locusts. And the locusts went up over all the land of Egypt, and rested in all the coasts of Egypt. Very grievous were they; before them there were no such locusts as they, neither after them shall be such, for they covered the face of the whole earth, so that the land was darkened. And they ate every herb of the land, and all the fruit of the trees which the hail had left and there remained not any green thing in the trees or in the herbs of the field, through all the land of Egypt. (Exodus 10:13–15)

In the summer of 2004, the sixth plague returned to the land. Children in Senegal, in western Africa, raced through the streets swarmed by insects. Bright pink, three inches across, these insects resembled thousands of tiny buzzing helicopters. Crops were ravaged. By December 2004, it was clear that another locust plague was definitely in the making. Swarms appeared in the northwest African countries of Morocco, Algeria, Mauritania, Senegal, Niger, and Mali. Trees in Morocco were covered with a wriggling pinkish gloss, made up of thousands of locusts. They devoured the leaves and sent a stream of droppings that sounded like rain to the ground below. The swarms covered hundreds of square miles and migrated thousands of miles, from the Atlantic to the Red Sea. By late 2004, they had already cost $122 million to control. We know that when locust swarming is at its peak, 20 percent of the earth's surface—from West Africa to the Middle East and Southeast Asia—can be covered with locusts. But perhaps this time we are better prepared than the pharaoh was to deal with them.

The desert locust, *Schistocerca gregaria*, is a unique insect with two distinct life phases. In most years, it is a quiet, solitary, dull-green grasshopper. It eats and flies at night, lives alone, and is a relatively unimportant inhabitant of much of western Africa. About every ten to twenty years, however, rains are sufficient to turn the desert green around the Red Sea and the Sahel, as they did in the summer of 2003. The Sahel region, south of the Sahara, was once considered semiarid but now has become a true desert because of overgrazing and tree clearing. When these breeding sites become lush and green, the hoppers reproduce rapidly, as each female lays several hundred eggs tucked under the surface of the soil. The eggs hatch after ten to sixty-five days, and over the next month or so, the locusts develop through five molts into adults. As the population grows with each generation, the juvenile hoppers come into closer and closer contact with each other.

Inevitably the rains cease. The land dries, and food for the hoppers becomes scarce. They begin to collect on the few remaining plants, touching each other. This touching, especially on their legs, leads them to change their behavior and, amazingly, their color as well. The juveniles change rapidly from dull green to bright yellow and black, a change so dramatic that not until 1921 did a Russian entomologist, Boris Uvarov, realize that they were both the same species. The black-and-yellow hoppers begin to collect together and move in "hopper bands" that become larger and larger as they travel across the ground, searching for more food. When they finally undergo their last molt to a winged adult, they take on a bright pink hue and become truly migratory locusts. They take to the wing and fly in masses for thousands of miles, carried by the wind.

Desert locust swarms occurred in Africa in 1988–89, after almost thirty years without such plagues. They appeared again in 1993–94 and 2004–2005. They are now known to breed in the Sahel, the Arabian Peninsula, India, and Pakistan. Although rain is certainly welcome in these semiarid regions, it is a mixed blessing. Unusually wet weather, followed by drying, is critical to development of the migratory phase. During the 1988–89 outbreak, an estimated 40 billion locusts infested forty-three countries. Three hundred million U.S. dollars were spent on insecticides to control the pest. In the end, though, this investment seemed to have been wasted; the insects flew out to sea on the wind, and many were destroyed. But now, the region contains more than twice the human population that it did in 1988. What will be the cost to those people from the next plague? And why have we not found a way to control these insects?

There are many reasons why the desert locust appears to come out of nowhere to overwhelm an entire region. They breed in remote areas, where there is poor communication and frequently civil strife. They fly long distances, often carried by the winds, and land in locations that are difficult to predict. The crops that they eat are very important to the local farmers but often do not have large commercial market value. This means that control measures, such as insecticides, may cost more than the local people can afford, and resources such as airplanes for spraying may not be available. Funding for control usually comes from international donors, who are much more likely to contribute to attacking a plague after it starts than to prevent it ahead of time. The locusts do not respect national borders, so coordinating control measures from one country to another is difficult. And finally, chemical insecticides have ecological and human impacts, such as killing beneficial insects, birds, and lizards, and they pose a threat to human health. These chemicals are not welcome in national parks and other ecologically sensitive areas and around the sources of human drinking water.

The British first recognized the desert locust as a serious pest during the 1920s in what is now Iraq. Conferences were held that led to the development in 1945 of an Anti-Locust Research Centre, which later became the Centre for Overseas Pest Research. This center was taken over by the Food and Agriculture Organization, a branch of the United Nations based in Rome, in 1953. Beginning in the 1950s, chemical insecticides were used in an attempt to control desert locusts, with little or no evidence that they stopped plagues. The first chemicals, organochlorides (related to DDT), were later found to be dangerous to humans and wildlife and were replaced by organophosphates and other insecticides. Now, twenty thousand tons of unused pesticides sit in these countries, posing a hazard to the communities.

Science often makes significant breakthroughs when scientists, governments, and funding agencies join together to attack a major worldwide problem. In 1989, the Centre for Agriculture and Biosciences International (CABI) in England and national and international organizations based in Benin, Mali, Niger, and Germany joined together to initiate a project called (in French) Lutte Biologique contre les Locustes et Sauteriaux (biological control of locusts and grasshoppers), or LUBILOSA. This organization received funding from the governments of Canada, the Netherlands, Switzerland, Britain, and the United States. LUBILOSA scientists began to look for an alternative to chemical insecticides for control of locust plagues. The logical candidates were diseases of the locusts themselves.

Schistocerca gregaria is the only grasshopper to undergo a phase change, but it is not the only grasshopper capable of migrating. In some regions of Africa, *Locusta migratoria* is a pest, while in Australia, the United States, and Canada, other species cause periodic damage to crops. These species became targets of the project as well.

The scientists focused on two major requirements: the disease had to be specific for grasshoppers and locusts, and it could not kill beneficial insects such as honeybees. They also recognized that the best candidates would be pathogens isolated from grasshoppers collected in the region where they wanted to use them. The hope was that these strains would be more resistant to the harsh environmental conditions of that part of the world.

Several pathogens had already been tested in the laboratory and the field against grasshoppers for many years. In 1911 a French scientist, Felix d'Herelle, claimed to have set off epidemics in grasshoppers in Mexico and South America using a bacterium, *Coccobacillus acridorum*, but his studies could never be reproduced. Another bacterium, *Serratia marcescens*, was tested in Kenya in 1959 with uncertain results. *Bacillus thuringiensis* products were not an option, because grasshoppers have acid digestive tracts and BT toxin crystals require alkaline conditions to dissolve. A protozoan parasite, *Nosema locustae*, had been tested for many years against grasshoppers in Montana and showed some promise, but it had to be produced in living insects, which did not seem practical for Africa. The LUBILOSA group screened nearly two hundred strains of pathogens isolated from grasshoppers around the world, and came up with several that looked promising. Among these were two fungus species that have roots in the history of insect pathology.

The first fungus to be tested against grasshoppers was the white muscardine fungus, *Beauveria bassiana*, first discovered by Agostino Maria Bassi in silkworms in the early 1800s. *Beauveria* was tested against grasshoppers and locusts in Canada and Mali. The white muscardine disease did efficiently kill the hoppers, but its activity was reduced at the high temperatures found in some of the most devastated regions of Africa and Asia. In cooler parts of the world, however, *Beauveria* could be effective as an insecticide. A strain isolated from grasshoppers in Montana was made into a product that was registered for use against rangeland grasshoppers in the United States in 1999.

But a second fungus, the green muscardine, *Metarhizium anisopliae*, looked even more promising. *Metarhizium anisopliae* was originally isolated from a beetle, the wheat cockchafer (*Anisoplia austriaca*) by Elie Metchnikoff in 1879. The infected insects sprouted green spores (from which the name green muscardine was derived). Metchnikoff named the fungus *Entomophthera anisopliae*. He found that it would efficiently kill two different species of beetles and suggested that this fungus might be useful as a microbial control agent against insect pests. The green muscardine fungus was moved to a different genus, *Metarhizium*, by the Russian L. Sorokin in 1887. Hundreds of strains of *Metarhizium* have been isolated from insects over the years since Metchnikoff's original discovery. Different strains of this fungus kill a wide range of insect hosts, including flies, caterpillars, grasshoppers, and many beetles (including the rhinoceros beetle, genus *Oryctes*). The activity of *Metarhizium* toward different species of insects varies greatly depending on the fungus strain. Strains that worked best against grasshoppers and locusts were isolated from locusts collected in two affected countries, Niger and Benin. These *Metarhizium* strains were found to infect only locusts and grasshoppers, making them attractive for the project.

Metarhizium has a simple life cycle, with none of the sexual or mating stages that are used to identify many fungi, so identification of different *Metarhizium* species depends on the size and shape of the spore. On the basis of these characteristics and modern techniques of genetics and biochemistry, the strains killing grasshoppers were called *Metarhizium anisopliae* variety *acridum*.

Metarhizium anisopliae var. *acridum* produces a furry mat of thin, hairlike fibers called hyphae on agar medium. When this culture exhausts the nutrients in the medium, it forms chains of oval spores (called conidia) that give the culture its characteristic dark green appearance. If one of these conidia lands on the outside of a grasshopper or locust under humid conditions, it germinates within sixteen to twenty hours. The spore sprouts a germ tube that may wander over the cuticle of the insect for several hours until it finds an appropriate site to penetrate. There it forms a flat plate called an appressorium. This structure sticks fast to the cuticle of the locust and begins to drill its way inside, using enzymes to dissolve the cuticle and mechanical force until it breaches the tough outer skeleton. Once inside, the fungus grows more hyphae and buds off bodies called blastospores that invade the whole body. The hopper becomes sluggish, stops eating, and eventually dies around ten days later. Once the fungus has used up all the nutrients in the body of its victim, it reverses its trek and pokes hyphal filaments back outside the grasshopper. On these it sprouts more green conidia spores, which can infect the next set of hoppers. Many scavengers, such as ants and birds, take advantage of the disabled hoppers and gobble them up. But if the decomposing body is left lying on the soil, infectious spores can remain for many months, waiting for the next victim.

The hopper is not without defenses against this enemy. We think that insects, considered to be "cold-blooded" animals, should have the same body temperature as the outside air, but the desert locust and other grasshoppers can set their own internal body temperatures by a process called thermoregulation. The locust does this by adjusting its posture to balance heat and cooling and moving from sun to shade during the day to keep a preferred internal temperature of around 38–40 degrees Celsius (approximately 100–102 degrees Fahrenheit). But when it becomes infected with *Metarhizium* or other pathogens, it can increase this internal temperature to as high as 42–44 degrees Celsius (approximately 108–111 degrees Fahrenheit). Locusts accomplish this "behavioral fever" by choos-

Grasshopper infected with *Metarhizium anisopliae*

ing warmer places to sit and basking in the sun. The maximum temperature at which *Metarhizium* can grow is approximately 37 degrees Celsius (99 degrees Fahrenheit). These thermoregulation and "behavioral fever" responses can greatly delay the progress of the disease. Scientists found that if infected locusts were not permitted to thermoregulate, more than 90 percent of them died in only ten days, but if they were permitted to increase their interior temperature, only 66 percent died over a period of seventy days. As researchers began to examine why this happens, they found that the number of blood cells remained high in thermoregulating locusts while infected locusts kept at a uniform cooler temperature lost many of these cells. As a result, the blastospores produced by the fungus multiplied rapidly in the cooler insects but were captured and killed by the blood cells of locusts held at warmer temperatures. Unusual proteins found in the blood also seemed to be associated with thermoregulation. These proteins may be an immune defense built up by the insect against pathogens.

Even the crowding behavior of the desert locust that leads to migration may have an effect on the ability of *Metarhizium* to kill its target. Although crowding among animals is usually associated with reduced ability to fight off disease, in the case of the locust, scientists were surprised to find that crowding increased its resistance to the fungus. These crowded insects were producing antimicrobial activity in their blood that was active against several different pathogens. The hopper, therefore, has many ways of getting around this potential biological control agent.

But *Metarhizium* also has weapons in its arsenal to overcome the grasshopper's defenses. It produces a set of small peptide toxins called destruxins that can kill the insect outright, and it may have other chemicals that can break down the insect's resistance mechanisms. It is better able to overcome the thermoregulating defenses of the locusts than is the other fungus candidate, *Beauveria*. Although *Metarhizium* does not grow well at the locust's feverish temperatures, it is not killed; eventually, when the locust has to return to cooler temperatures at night, the fungus may win the battle. And if it infects early in the life of the hopper, it often reduces the number of eggs that are laid. The fungus-grasshopper interaction has turned out to be complicated indeed.

In the past, fungi have often looked like promising biological control agents, but have proven notoriously difficult to get to work well in field situations. They tend to lose activity rapidly if exposed to sunlight and heat and are effective pathogens only when the humidity is high. Many of the targeted locust regions in Africa experience high temperatures, lots of sunlight, and little rain, which are not exactly optimal conditions for fungus-based insecticides. In addition, many people are allergic to fungi or molds, which is another cause for concern in considering widespread use of these organisms in biological control efforts.

Metarhizium does have many useful aspects, though. It has a simple life cycle that makes it easy to culture in the laboratory or a commercial factory. It can be easily produced in large quantities on cheap artificial media, some as simple as sterile grain. Its spores penetrate the cuticle of the hopper and multiply inside the insect, reducing the feeding and egg-laying capabilities of the infected host. After the hopper dies, spores may be produced inside the insect that can persist through the dry season to infect the next generation of insects. And it kills relatively quickly. Above all, it has no effect on nontarget organisms—including birds, fish, mammals, and beneficial insects—and is not a hazard to humans.

The collaborative LUBILOSA group then tackled the next challenge: how to make *Metarhizium anisopliae* var. *acridum* into a useful biological control agent for locust-stricken regions of the world. The scientists first produced large numbers of spores and dried them down with the idea of shipping them to plague areas when they were needed. Alas, the LUBILOSA collaborators soon found that the spores had lost their ability to germinate within a few weeks. The spores were absorbing moisture from the atmosphere, attempting to germinate, and then dying. If the temperature was too high, it killed the spores. The researchers also needed to solve the problem of the spores being killed by ultraviolet radiation from the sun, which is a real concern in semidesert areas of Africa. Although some strains appeared to be better than others at resisting UV, no more than 35

percent survived even four hours of simulated sunlight in the laboratory. But in the field, some survived better. The scientists had to
find a way to protect the spores from the sun and to prevent them
from absorbing moisture.

It was not easy to find a company willing to experiment with
mass production of a fungus to kill insects in poor countries where
the company would gain little or no profit. Finally the LUBILOSA
group constructed its own production plant in Benin. Workers there
first grew the fungus in a liquid broth medium consisting of simple
sugar and yeast. After the fungus had achieved high numbers in
the broth, it was applied to sterile rice held in plastic bags. There it
grew and produced the characteristic green spores. When sporula-
tion was complete, the rice was allowed to dry for about two weeks.
After considerable research, a dry product was produced that could
survive for up to four years when kept at the proper temperature
and humidity. This first product became essential for the development of an insecticide that could be used in the field.

The target areas for desert locust control are thousands of square
miles in size, and often the only way to efficiently treat these remote areas is by airplanes or trucks with high-pressure spraying
equipment. Such an application method requires insecticides to be
sprayed in "ultra-low volume," that is, small amounts are spread like
a fog over a wide area. The holes in the nozzles of these sprayers
are quite small—not much larger than the *Metarhizium* spores—and
can clog up very easily. The *Metarhizium* spores had to be harvested
carefully from the rice grains, leaving no residue in the product that
would clog the equipment. These sprayers require a liquid preparation, not a dry powder, and water is often scarce in the desert
regions where locusts breed, so oil formulations made more sense.
Vegetable- and mineral-oil-based formulations were developed by
the LUBILOSA group in Benin and were found to have several
other advantages as well. For one thing, they tend to stick to the
outside cuticle of the hopper, increasing the chances that the insect
will become infected. The oil also seems to protect the spores from
ultraviolet radiation and moisture, so they remain active for much
longer, even surviving between seasons.

The team took the experimental product to the field in 1995–97 and achieved encouraging results. Compared with chemical insecticides, which killed the hoppers immediately, the fungus took two to three weeks to kill. But the locusts soon returned to the chemically treated sites, whereas in fungus-treated sites, their numbers continued to decline for many weeks. Nontarget beneficial insects were not affected by *Metarhizium*, but they were definitely reduced by the chemicals. The efforts of the scientists paid off. A new locust control product based on *Metarhizium anisopliae* var. *acridum* was ready to be licensed to a commercial company. The registered product was named Green Muscle.

A product made in South Africa became available in 1998. African countries affected by the desert locust have rapidly moved to use Green Muscle, and the UN Food and Agriculture Organization recommended it. Niger began to use this product in its grasshopper control operations in 2000, becoming the first country to do so. Niger sprayed 300,000 hectares (750,000 acres) in 2004 and 2005 with this fungus. Governmental and nongovernmental organizations have given financial support for its use. Farmers are learning that the slow kill of locusts and grasshoppers does not mean that a product will not work, and they appreciate that their homes, water, and food sources are not being sprayed with chemicals.

The locust has been a plague in Africa, Arabia, and Asia for thousands of years. The Arabic word for the desert locust is translated "teeth of the wind." But this time, a fungus disease may pull some of its teeth.

14

Ad Infinitum
Insect Symbiotic Organisms

All living things require the same basic nutrients to survive and reproduce: water; carbohydrates such as sugars; amino acids to form proteins; and small amounts of vitamins and minerals. Plants and other photosynthetic organisms manufacture many of these nutrients for themselves, while animals acquire most of these substances from their food. This is why we humans must eat at a minimum a wide variety of plant products, which we then convert into the substance of our own bodies. If, however, you were confined to a forest, with nothing to eat but tree leaves and wood, you would eventually perish. But millions of insects survive quite nicely on specialized diets consisting of a single thing: leaves, wood, plant sap, or even blood. How can a termite or a bark beetle spend its entire lifetime gnawing wood? How does an aphid or whitefly reproduce rapidly while eating nothing but plant sap? How do cockroaches proliferate while eating dry leaves and the detritus from our kitchen floors? And how do lice and fleas survive by feeding on nothing but blood?

The answer to this question lies in the term *symbiosis*. All of these insects, and many more, coexist with an amazing array of microorganisms that provide part or all of their nutritional requirements.

In return, the insects provide the raw materials needed for reproduction and growth of the microorganisms and then share in the nutritional rewards. Many different types of bacteria, fungi, and protozoa are associated with insects, and these associations date back millions of years. Some of the symbionts, as the microorganisms are called, cannot live outside the body of the host insect. These organisms have relied on their hosts for so long that they have lost critical genes necessary for independent living. And some of the insects likewise cannot survive and reproduce without their symbiotic organisms.

The term *symbiosis* has been defined in many different ways. The German botanists Heinrich Anton deBary and Albert Bernhard independently developed the concept of symbiosis in 1868, and deBary coined the term *symbiosis* in 1879. He used it to refer to the situation in which two dissimilar organisms are living together. By deBary's definition, symbiosis includes a very complex set of associations, ranging from accidental relationships between two (or more) organisms with no benefit to either (e.g., the bird living in your backyard), to full parasitism, in which one entity benefits and the other is harmed (e.g., fleas on your dog). In between these extremes are relationships that benefit one but do no harm to the other (e.g., flies eating your garbage), and relationships that give equal benefit to both partners (e.g., cows and farmers). In the examples used here, you can easily imagine ways in which the relationship can shift from benefit to harm—if the flies carry disease, for example, or if the farmer decides to slaughter the cow. The farmer also has control over another critical factor in the life of the cow, her reproduction. But if the farmer is very poor and has only one cow, then he and his family also depend upon the cow. Insects and microorganisms are involved in all these sorts of relationships, and many more.

A key player in our understanding of these complex interactions was another German scientist, Paul Buchner. Buchner, the son of a physician, grew up in Nürnberg, Germany, and took a great interest in botany. At the tender age of nineteen, while still in gymnasium (a high school for university-bound students), he published two articles on how certain aquatic insects construct their protective cases. Although he enrolled in the University of Würzburg in 1907 intending to become a botanist, lectures by prominent zoologists lured him back to the study of insects, and he proceeded to obtain a doctorate in this field. Buchner then took a position at Munich, where he began to study microorganisms that he could see within the bodies of plant-sucking insects. Beginning in 1911, Buchner studied these organisms in dozens of different insects, a fascination that would last over fifty years. He discovered what appeared to be special organs in aphids, cicadas, and whiteflies, filled with bacteria-like organisms. These organs seemed to be transmitted from the mother to the offspring directly, often through the egg. Some of the insects even appeared to have several different kinds of microorganisms within their bodies. In 1921 he assembled everything that was known at the time into a seminal book, *Endosymbiose der Tier mit pflanzlichen Mikroorganismen* (*Endosymbiosis of Insects with Plant Microorganisms*). He revised this book four times, the last in 1965 in English. In the dedication of a book to Buchner in 1967, Anton Koch stated that Buchner's lifetime of work in the field of insect-microbial symbiosis "gave us the key to the magic garden of symbiosis."

"Magic garden" is an appropriate term for the way in which leaf-cutter ants, wood-boring beetles, and some termites make a living. Each of these insects relies on "ambrosia," the food that gave immortality to ancient Greek gods. How shocked the Greek gods would have been to learn that ambrosia was actually fungus cultured by insects!

In 1836 J. Schmidberger observed winding black trails, resembling a road map, under the bark of a tree. Inside these trails was a white substance that he thought was a mixture of insect spittle and sap, and he called it ambrosia. But he was only partly correct. What he saw was the result of an insect at work, but the ambrosia was actually a fungus. In each of these trails, an ambrosia beetle of the genus *Xyloborus* had raised a family.

The female ambrosia beetle is diploid, that is, she comes from a fertilized egg. Males, by contrast, come from unfertilized eggs (and thus are haploid) and cannot fly. Males are rare — as more than twenty to thirty females may be produced for every male — and the male's only function is to fertilize eggs. The mated female finds an appropriate tree, burrows beneath the bark, and establishes a fungus garden from spores carried in a special pouch at the base of her mandibles or in her thorax. The female beetle carefully tends her fungus garden and her offspring until they emerge as adults, when she usually dies in her nest. If the female should die prematurely, the fungus will take over and kill the offspring. The constant grooming action of the mother beetle converts the wild form of the fungus, which can be fatal to the offspring, into the ambrosia form, which they eat. This symbiotic relationship is one that can quickly slide from mutual benefit to harm for the host beetle.

Leaf-cutter ants in the genera *Atta* and *Acromyrmex* are confined to the Americas, where they are commonly seen running along the forest floor carrying large pieces of leaves, flower parts, or woody fruits. After her mating flight, the queen leaf-cutter ant carries a bit of fungus from her mother's garden in a special pocket in her mouth. She digs a small cavity in the soil, ejects the fungus pellet, and fertilizes it by defecating on it. There she lays her first eggs and starts the colony. The ant colony eventually burrows many feet into the soil and consists of thousands of ants, including workers, caregivers for larvae and the queen, gatherers, and soldiers. The gatherer ants carry the leaf sections like parasols over their heads, running on distinct trails back and forth between trees and their nests. In the process of cutting leaves, these ants can be very damaging to crops and ornamental plants. Each piece of leaf is carried underground into a large nest with many open chambers where worker ants cut up the leaves into small fragments. They push these into a white garden of fungus hanging from the ceiling and walls of the chambers. The fungus rapidly grows over the leaf and consumes it, producing swollen cells that provide food for the ants. In turn, the fungus makes enzymes that end up in the fecal fluid of the ants. This fluid is applied to the plant fragments, helping to degrade the tough leaves into a form that the fungus can utilize.

As the queen ant lays each egg, it is removed by a caregiver and placed into the fungus garden, where it hatches and is fed fungus by

a sister ant. The entire colony relies totally on the fungus for food.
Nearly two hundred different species of leaf-cutter ants rely on an
equally complex set of fungal species. The leaf-cutter ant/fungus collaboration is a very efficient factory for the production of both ants and fungus.

Termite mounds are one of the most memorable sights on an African safari. These cone-shaped mounds can stand taller than a giraffe and are made of hard, bricklike mud. For every aboveground foot of mound, many more feet extend underground, filled with termites. These mounds are busy cities of termite activity. Mound-building termites of the Macrotermitinae group, which are found only in Africa and South Asia, have important economic effects on local inhabitants, as they are destructive to crops such as wheat and sugarcane, utilize nutrients in forests and orchards, and rapidly consume entire wooden buildings. The mounds also make plowing and planting very difficult. (However, mushrooms emerging from the mounds during the rainy season are a prized delicacy in Asia.)

All termite species live in social colonies founded by a pair of reproductive adults (the king and queen), which produce offspring workers and soldiers. They feed on wood or wood products, and many rely on bacteria and protozoa in their guts to digest cellulose and lignin for them.

The Macrotermitinae in the Old World are unique in culturing large fungus gardens in above- and belowground mounds. The king and queen in this colony mate and then seal themselves forever in a hard clay vault belowground. There they pull off their wings and establish their nest. The queen's abdomen becomes huge and sausagelike, many times the size of her modest mate and her offspring. She becomes an egg-laying machine, turning out a new egg every few seconds. The first offspring become workers that feed the parents, and as the colony grows, workers remove the eggs to the fungus garden. There they deposit the eggs under the garden, where the new nymphs will feed and grow. Workers tirelessly care for the chambers, lick the queen, repair cracks in the walls, feed the queen and king, and groom the fungus garden. Soldiers defend the mound by squirting sticky, irritating substances on any potential attackers.

At least two species of fungi have been found in these mounds, and these species often appear to coexist. Their function is to decompose the wood and leaf litter collected by workers and to convert it into food that the termites can digest. In payment, termites ensure a perfect home for the fungus and abundant plant material. This relationship is a symbiosis in the classical sense, in which both members benefit.

The intimate relationships between fungi and termites, ants, and beetles are only a few examples of the complex relationships helping to provide nourishment for both insects and symbiotic microorganisms. Many different types of bacteria, protozoa, and fungi live in or on insect bodies. Some insects, such as the tsetse fly (carrier of African sleeping sickness) have specialized bacteria attached to their gut cells that provide necessary vitamins and other nutrients. Without these, the flies cannot reproduce. In some termites, gut bacteria fix nitrogen out of the air and contribute to nutrition of the insect. Bacteria may even produce sex attractants that allow insects to find a mate. Protozoa are also found intimately associated with the gut cells of some termites, helping them to digest wood. The interaction of insects with microorganisms in their digestive tracts is a complicated, very old relationship.

Two of our least favorite insects, the louse and the cockroach, were players in the discovery that bacteria can live not only in the guts of insects, but also completely within their cells. Robert Hooke, using one of the earliest microscopes, saw a strange organ in the human louse in 1665. He thought that this might be the "liver" of the louse, but he was unable to tell what was inside. As good microscopes became widely available in the 1800s, scientists began to see many more such objects within insects. In 1879 deBary saw cells in insects that he called mycetocytes inside special organs that he called mycetomes. The objects inside the mycetocytes were bacteria-like bacteroids. We still use deBary's terms today. Elie Metchnikoff, in addition to his early studies of immunology, observed mycetomes in aphids, and he thought that these mycetomes were probably supplying nutrition to the embryos. It was F. Blochmann's observations of objects in the eggs, fat bodies, and embryos of cockroaches that con-

vinced him that these objects actually were bacteria. Blochmann's
publications in 1887 and 1888, describing these objects and his attempts to culture them from cockroaches, were seminal to the understanding that microorganisms could live within special organs inside certain insects. These organs were later called Blockmann bodies.

By the early 1900s, scientists were attempting to culture the organisms within insect mycetomes and even to transfer the organs from one insect to another. These experiments met with mixed success. In some cases, apparently successful bacteroid cultures were later found to be those of contaminating bacteria. However, as more sophisticated technology and antibiotics became available, we began to understand the complexity of the associations between insects and these intracellular symbionts.

Following Blochmann's critical discovery of intracellular bodies in one species of cockroach, these objects were eventually found in

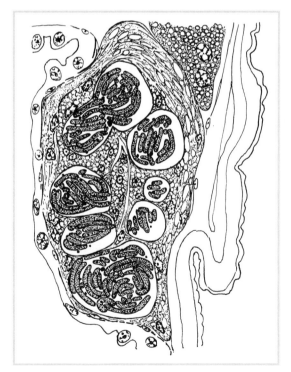

Symbiotic bacteria in the gut of a louse (drawing by G. Buchner)

nearly all species of roaches. The mycetocytes are usually in the fat body, an organ that has several functions in insects, including production of hormones and storage of nutrients. The fat body takes up much of the abdomen of the roach and is the smelly white substance that squirts out of a roach when you step on it. The mycetocytes enter the ovaries in females and are transmitted into the egg. When the egg is fertilized and the embryo begins to grow, the mycetocytes migrate into the embryo where they enter the developing fat body, a process called vertical transmission.

Scientists eventually began to ask whether these bacteria-like organisms are really bacteria, and whether they are essential to the insect. In the mid 1950s, Marian Brooks and Glenn Richards, at the University of Minnesota, fed German cockroaches a diet that contained several types of antibiotics. To their surprise, they found no differences in the life spans of the treated roaches and those that still had their bacteroids intact. However, the offspring of roaches fed the antibiotic aureomycin were smaller than normal and were a different color. These newly hatched roaches were feeble; many died immediately after hatching. Upon examining these insects under a microscope, Brooks and Richards found that they completely lacked the bacteroids. Therefore, the antibiotic appeared to interfere not with the bacteroids in the mother and father, but rather with transmission of the microorganisms to the egg. As Brooks and Richards began to experiment with diets for these insects, they found that a high level of yeast in the diet helped the roaches to grow without their symbiotic organisms, but the offspring of treated roaches still took two or three times as long as those of healthy roaches to mature. The yeast apparently added nutrients that normally would have been provided by the bacteroids. In later experiments, Brooks and Richards injected offspring of antibiotic-treated roaches with bacteria or entire mycetomes from a healthy roach. The transplantation experiment worked; the grafted tissues grew and apparently provided necessary nutrients to the insects. But they did not enter the ovaries, so the insects still did not produce normal offspring. In 1958 Richards and Brooks suggested that these microorganisms contributed nutrients to the insect, and this served as the first clear explanation of why this primitive insect, the cockroach, should have symbiotic organisms. This partnership is a true symbiosis; neither the roach nor the bacteroid can long survive on its own.

Aphids, whiteflies, leafhoppers, and scale insects spend their entire lifetime with beaks plugged into sap channels, the circulatory system of a plant that moves water and photosynthetic products such as sugar through the plant. A diet of plant sap is the equivalent of surviving on nothing but soda pop because sap contains sugar and water but little else. In particular, plant sap is very short in the amino acids that are absolutely necessary for building proteins in the body of the insect. Once again, intimate relationships with intracellular microorganisms allow these insects to survive on such a diet and to reproduce at remarkable rates.

More than 200 million years ago, a microorganism infected aphids and co-evolved with the insect. Now, these two cannot survive separately. Inside each aphid are large mycetomes composed of mycetocytes full of spherical or oval microorganisms. Alongside these may be rod-shaped "secondary" symbionts as well. All these organisms are passed from the mother aphid to her offspring through the egg. U. Pierantoni and K. Sulk first described symbiotic organisms in organs within the aphid in 1910. The genus of these symbionts was later named *Buchnera* in honor of Paul Buchner's many studies of symbionts.

Aphids are major pests of agricultural and ornamental plants. Millions of tiny wingless aphids can live on an acre of crops, and their sucking activities weaken the plants. They can also transmit lethal viruses to the plant. The aphid inserts her beak into a plant's sap channel and secretes a sheath around her beak. With this "soda straw," she pumps sap out of the plant, absorbing the small amounts of nutrients and excreting the sugar and water as honeydew. This sticky substance contaminates the leaves and flowers. During the summer, many aphids produce live offspring without mating. Inside each female aphid is a female offspring, and inside this offspring is another female offspring. Female wingless aphids undergo live birth, which explains how aphid populations can explode practically over-night. In the fall, winged males and females are produced, mate, and lay eggs that survive over the winter.

At each of these steps in reproduction, *Buchnera* is passed to the next generation. Each aphid may carry a hundred mycetocyte cells, each containing thousands of *Buchnera*. The *Buchnera* microorganisms must be passed from the mother encased within their protective mycetocyte cells. If free *Buchnera* cells are injected into an aphid, they are rapidly destroyed by the insect's immune cells, which treat them as foreign objects.

For many years, scientists thought that the *Buchnera* microorganism was related to rickettsia or mycoplasma, intracellular organisms that, in humans, cause diseases such as Rocky Mountain spotted fever and some types of pneumonia. However, as new genetic tools became available in the early 1990s, Paul and Linda Baumann and their colleagues at the University of California discovered that *Buchnera* is really a bacterium, related to our familiar gut bacterium, *Escherichia coli*. This led them to name this organism as a new genus and species, *Buchnera aphidicola*. Angela Douglas at the University of York in England and other scientists have fed aphids antibiotics to kill the microorganisms, and then the researchers have analyzed the aphid bodies to learn what was reduced or missing. These studies have shown that *Buchnera* is the amino acid factory for the aphid. Thanks to the efforts of scientists from Japan (Hajime Ishikawa at the University of Tokyo) and the United States (Nancy Moran at the University of Arizona, researchers at the Baumann laboratory, and several others), the genes of *Buchnera* have been examined in great detail. Now we understand a great deal more about why *Buchnera* is totally dependent on the aphid, and why the aphid is dependent on *Buchnera*. The two players trade complex metabolic products, and neither has all the genes necessary to survive. Insects cannot make ten "essential" amino acids (most of which are not present in plant sap) themselves, but *Buchnera* can synthesize some or all of these amino acids. In return, *Buchnera* gets a stable, nutrient-rich environment and is passed along to the next generation of aphids. It no longer has any other way of moving from one insect to another.

Amino acid production by *Buchnera* clearly is the key to the aphid's success, which confers on it the ability to be a major pest. *Buchnera* is a slave of the aphid, and the aphid is a slave to *Buchnera*.

As we examine the list of microorganisms that have struck up symbiotic relationships with insects, we find that one group appears to be missing. There are bacteria, fungi, and protozoa in these associations, but what about viruses? Certain species of parasitic wasps called ichneumonids and braconids are perhaps the only animals known to have a symbiotic relationship with a virus. Unlike the melanization of a wasp larva observed by Metalnikov in 1919, when these wasps inject their eggs into caterpillars, the caterpillar does not respond with a black melanized cluster of cells. Somehow the wasp has fooled the caterpillar into accepting the egg. In the 1970s, scientists discovered that small particles that resembled viruses were being injected along with the wasp eggs. These particles overcame the immune response and also prevented the caterpillar from changing into a pupa and then into a moth. The caterpillar became a factory for production of the next generation of wasps. The ability of a virus to shut down the immune system of the host is not unknown; this is the way that the human immunodeficiency virus causes the disease AIDS. So what were these wasp-carried viruses?

At first they were assumed to be baculoviruses, relatives of the nuclear polyhedrosis virus, but as the DNA of the wasp virus was studied further, the virus was placed into a separate group named polydnaviruses. When the particles injected along with the wasp eggs were examined closely, researchers found that the particles could not multiply in the caterpillar because they are missing some vital genes. They replicate only in the ovaries of the wasp and then go out with the egg to overcome the immune and developmental systems of the caterpillar. In some wasp species, the venom injected along with the egg and virus enhances this immune-response-system shutdown. Even more surprising is the finding that much of the virus DNA may have actually originated in the wasp's own genes. This relationship is obviously extremely old, estimated to have originated around 65 to 75 million years ago, back in the days of the dinosaurs.

As we well know, mosquitoes carry many dangerous diseases

such as malaria, yellow fever, dengue, encephalitis, and West Nile fever. For this reason, mosquitoes have been the subject of many scientific experiments. To carry out these experiments, scientists had to develop strains of mosquitoes that will survive and reproduce quickly in the laboratory. While working with strains of the encephalitis and West Nile vector, *Culex pipiens*, in the 1950s, S. Ghelelovitch and S. Laven made an extraordinary discovery. Certain strains of *C. pipiens* collected from different geographic regions would cross with each other and produce live offspring only in one direction—that is, males from one strain crossed with females of the second strain would produce offspring, but females from the first strain would not produce offspring with males of the second. Males and females within a strain would always produce offspring. If these insects were all the same species, how could this be? By definition, animals of the same species should be able to mate and produce viable offspring. Laven was able to show that the female mosquito appeared to be controlling whether she produced offspring with a male. This substance seemed to be in the cytoplasm of the egg. Laven called the situation "cytoplasmic incompatibility."

Thirty years earlier, in 1924, M. Hertig and S. B. Wolbach had observed rickettsia-like microorganisms in the reproductive tissues of some strains of *Culex pipiens*. These organisms were later named *Wolbachia pipientis* by Hertig in 1936. But the connection between *Wolbachia* and cytoplasmic incompatibility in *Culex* mosquitoes was not made until 1971, when J. H. Yen and A. Ralph Barr produced a new hypothesis connecting these observations. Yen and Barr fed antibiotics to some of the *Wolbachia*-infected female mosquitoes and then attempted crosses with males of their own strain and males of strains without *Wolbachia*. Suddenly, these females could no longer produce offspring with males of their own strain, but now males of a noncompatible strain proved fertile. Thus *Wolbachia* appeared to be controlling the reproduction of the mosquito. Soon cytoplasmic incompatibility was discovered in many species of insects, including flour beetles, alfalfa weevils, planthoppers, flour moths, parasitic wasps, and the fruit fly, genus *Drosophila*. *Wolbachia* was not the only organism with these controlling abilities; other bacteria and

even protozoa were involved, and sometimes more than one strain of
Wolbachia was present. Not only do these organisms control which strain a female can successfully mate with, but they can also skew the sex ratio of the offspring toward females and sometimes totally eliminate male offspring.

The critical role in the mosquito story appears to be the male's sperm. If *Wolbachia* is present in the male, and he mates with a female with the wrong strain of *Wolbachia*, the egg may go through one or two divisions but then development stops. But strangely, the microorganism cannot be found in the mature sperm, the effects appear to be somehow exerted on immature sperm and then carried over to the egg. If the female contains the same *Wolbachia* strain, fertilization is successful.

Wolbachia has many tricks up its sleeve. It also determines the sex of an isopod crustacean, a distant relative of insects. We usually think of sex as determined by X and Y chromosomes and the sex ratio as approximately half male and half female. But some females of the wood louse, genus *Armadillidium*, routinely produce nearly all female offspring. In 1973 a group of French scientists led by G. Martin discovered that this odd ratio is related to the presence of small bacteria found in the cells of some females but never in males. Molecular techniques have confirmed that these bacteria are *Wolbachia*. When young *Armadillidium* females were held at warm temperatures or fed antibiotics that killed the bacteria, the females gradually turned into males! In fact, all the females turned out to be males that had had their sex reversed by the presence of the bacteria. If by chance an embryo missed acquiring *Wolbachia* from its mother, it became a male; otherwise all her offspring were daughters. This total control over the sex of the host appears to be restricted to crustaceans only.

Another *Wolbachia* trick was hidden from view until quite recently. Parasitic wasps are very useful biological control agents and have been the subject of intense research to develop ways to rear them in large quantities for release in agricultural areas and greenhouses to control pests such as aphids. Several species of these wasps are sold commercially as biological control agents. While rearing some of these wasps, scientists discovered that the females appeared to be producing only daughters and that they reproduced without mating. This is a very desirable trait, as only the females are useful in biological control, since they alone parasitize and kill the pest insects. However, during the summer or when temperatures in the lab were allowed to rise above a certain level, males suddenly appeared! Studies during the 1960s showed that the temperature to which the female is exposed during development and adulthood determines whether she produces all daughters, or some sons as well. It also determines how many sons she might produce.

Up until this time, scientists assumed that the ability to produce daughters from unfertilized eggs was based on genes in the nucleus of the wasp's cells. Using a complicated set of mating experiments, a group of scientists led by Richard Stouthamer determined that genes actually are not involved in determining the sex of the offspring of a tiny parasitic wasp, *Trichogramma pretiosum*. They also found that certain antibiotics as well as higher temperatures could lead to the production of male offspring. This led them to search for microorganisms; they soon found these microorganisms in the eggs of female wasps that produced only female offspring. Once again, the culprit was *Wolbachia*.

Strains of *Wolbachia* that can induce the production of daughters from unfertilized eggs (the process called parthenogenesis) occur in many parasitic wasps. But how can this process work? We know that bees produce offspring from unfertilized eggs, but these are haploid males, having only one set of chromosomes from the mother. The wasp produces daughters that are diploid, with two sets of chromosomes, from an unfertilized egg. Researchers eventually discovered that meiosis, the cell-division cycle that leads to haploid eggs and sperm, is altered at the first step of the cycle, resulting in single chromosomes becoming double. But if a male is around and mates with a female wasp, the sperm will be used to fertilize an egg producing a diploid daughter. Somehow the parthenogenetic process is bypassed by normal mating, permitting occasional genetic exchanges. A complicated system indeed!

There are real economic benefits to using all female, parthenogenetic wasps for biological control. For example, they can be mass

reared, and all of the wasps (not just some of them, as is the case when males are also produced) will be useful in killing the target pest. Also, parthenogenetic wasps — all egg-laying females — should outcompete strains that produce lots of nonparasitizing males. These should be desirable strains to release in the field. But this assumes that *Wolbachia* does not take a toll in energy and reproductive capability of the wasp, which it has been shown to do in some wasps. Sexual reproduction does have some advantages. Genetic exchange allows bad genes to be removed from the population, and good ones to be moved around. The wasp and *Wolbachia* thus appear to be involved in a fascinating balancing act.

Another important question is, how does *Wolbachia* benefit by controlling the reproduction of its host, whether by cytoplasmic incompatibility, by producing only females, or by parthenogenesis? The most obvious benefit is that *Wolbachia* thereby ensures that its offspring are passed to as many host offspring as possible. Since this takes place through the egg and not the sperm, it makes sense for *Wolbachia* to control who mosquitoes mate with, what sex a wood louse becomes, and what sex of offspring a wasp can produce. *Wolbachia* is obviously very successful, as almost 20 percent of all insects, representing all major taxonomic groups, are infected with this organism.

We can also ask, does the insect benefit from this relationship? Some of the effects of *Wolbachia*, such as increasing the female-to-male ratio in the offspring, are beneficial to the host. Additionally, *Wolbachia* may indeed contribute something essential to the host, as treatment of some wasp species with antibiotics to cure them of their symbionts leads to reduced reproduction, but in other wasps exactly the opposite occurs, so this aspect of the relationship is not yet well understood. And the peculiar effects of *Wolbachia* on the reproduction of its hosts, as in cytoplasmic incompatibility, can decrease the success of insects that are not infected. These questions have been the subject of intensive study and mathematical modeling. The relationships between *Wolbachia* and its insect hosts thus may be beneficial, neutral, or harmful, depending on a plethora of factors.

Clever scientists have recently attempted to take advantage of the tendency of *Wolbachia* to move through an insect population by introducing certain genes into *Wolbachia* with the goal of reducing the harm from these insects. In particular, research has focused on insects that vector human diseases. In the 1960s, a successful operation was undertaken to eradicate the screwworm fly (*Cochliomyia hominivorax*) from Mexico and the southwestern United States using males sterilized by radiation. These males mated with fertile females and produced no offspring, and the fly was brought under control. Following this successful model, some researchers wondered, why not use *Wolbachia* in a similar way to control reproduction in mosquitoes? The World Health Organization undertook such a trial in the 1960s. Cytoplasmically incompatible *Culex* mosquitoes imported from France and the United States were introduced into small isolated mosquito populations in Burma. This experiment was a success, as none of the eggs produced within the next twelve weeks hatched. In other experiments, tsetse flies (carriers of African sleeping sickness) have been targets of this program. Males containing the appropriate *Wolbachia* strain are chosen to be incompatible with local females and are released into the population. Unfortunately, at this time, moving from these small-scale experiments to large-scale applications in developing countries seems unrealistic.

Wolbachia has a tendency to move rapidly through mosquito populations, making it a very attractive candidate for manipulation of vector populations. Scientists are beginning to ask, is it possible to move genetically engineered *Wolbachia* or other symbiotic organisms into disease-causing insects, with the goal of reducing the host insect's ability to transmit human disease? For example, could mosquitoes with altered symbionts be induced to die before they can transmit malaria? More than 12 million people are infected with Chagas disease, carried by kissing bugs, and millions more are infected by African sleeping sickness, carried by the tsetse fly. Could symbiotic bacteria of kissing bugs or tsetse flies be genetically engineered to kill the disease-causing trypanosomes carried in their gut? Many of these ideas are currently under development. In 2004, the entire *Wolbachia pipientis* genome was sequenced, a step that will un-

doubtedly lead to future discoveries and uses for this organism. The
ability to manipulate insect symbionts, just as these organisms manipulate the insects, leads to intriguing future possibilities for control of life-threatening insect-vectored human diseases and agricultural pests.

As we begin to understand more about the complex interactions
between insects and their symbiotic microorganisms, these interactions do indeed seem to go on ad infinitum.

15

Postscript

Landmarks
That Tie
Our Stories
Together

As you have read these stories, you will have noticed that major advances in different fields often occurred at around the same time. For example, the basic understanding that infections cause diseases of insects and humans really began in both Europe and Asia around 1850–1900. The first genes from both baculoviruses and *Bacillus thuringiensis* were cloned within the decade from 1972 to 1982. The timeline that follows illustrates that the correlation of these events was no accident. In any field of science, discoveries and developments outside the specific field may have an important impact on research, sending it off in new and exciting directions. The first good microscopes permitted scientists to see a marvelous new world of microorganisms, including diseases of invertebrates. The concept that microorganisms can cause disease—put forth by Koch, Pasteur, and others during the second half of the 1800s—profoundly influenced scientists fascinated with diseases of invertebrates as well. The ability of Pasteur to immunize children against rabies certainly influenced Metchnikoff and others to ask whether invertebrates could be immunized.

Many of the developments in science do not really depend on dis-

coveries, as you may have been led to believe in school, but rather
on technological developments—machinery, tools, chemicals, laboratory equipment, and techniques. Although these developments usually have originated in a scientist's mind and have been explored in a laboratory setting, their great utility depends upon their commercial production and widespread availability. Even simple items such as inexpensive, sterile plastic pipettes, test tubes, bottles, and petri dishes profoundly enhance the amount of work that a scientist can accomplish in a day. The development of complex, expensive equipment such as the electron microscope in the 1930s and 1940s and the polymerase chain-reaction machine, used to sequence genes, in 1985 led to rapid advances in our understanding of the agents of disease, their genes, and mechanisms for both controlling human disease and using invertebrate pathogens for beneficial purposes. The development of the computer, its use in analysis of chemical structure and DNA sequences, the assembly of huge computer DNA databases, and even the rapid communication among scientists that the Internet provides have had a profound impact on scientific research. Prior to the Internet, dissemination of scientific findings required months or longer; it now may take place in a matter of minutes. Readily available, affordable international air travel has led to collaborations among scientists from around the world and permits them to attend meetings at which they can discuss findings and ideas with colleagues from many countries and fields. International collaboration has been a hallmark of invertebrate pathology since the first International Colloquium on Insect Pathology and Biological Control held in Prague in 1958.

An examination of the following timeline should give you a sense of how the field of invertebrate pathology progresses hand in hand with developments in other areas of biology, chemistry, and physics. It provides good examples of how sharing of ideas and technology among fields can lead to rapid advances in science. And it leads us to ask, what will be next?

Events described in the text are typographically
distinguished by this typeface.

Before 1500

330 BC   **Aristotle describes diseases of bees**

AD 555   **silkworm eggs smuggled into Europe**

1498   **Europeans travel to India, bringing back cholera**

1500–1800

1527   **Vida describes diseases of silkworms in Italy**

1530   Fracastoro suggests that syphilis is transmitted by "seeds"

1629   Harvey describes human circulation

1662, 1666   first scientific societies established (Royal Society of
              London; Academy of Science, Paris)

1665   first scientific journal published (*Philosophical Transactions
          of the Royal Society of London*)

1679   **Marian describes diseases of silkworms in Germany**

1683   van Leeuwenhoek observes "animalcules" in a horsefly

1752   first peer review of scientific articles (*Philosophical
          Transactions of the Royal Society of London*)

1771   **Schirack describes diseases of bees**

1796   Jenner uses cowpox to vaccinate against smallpox

1800–1860

1817   **first cholera pandemic**

1830s   development of good microscope lenses and the
           compound microscope

1834   **Bassi describes fungus disease of silkworms (first disease
          confirmed to be caused by a microorganism)**

1838   Schwann and Scheiden propose that all living things are
          composed of cells

1848   Semmelweis introduces antiseptic techniques

1854   Snow relates contamination of wells with sewage to
          cholera; Pacini describes the bacterium causing cholera

1854   Eastman develops the Kodak camera

1856   **Cornalia and Maestri see crystals in nuclei of silkworms
          with "jaundice"**

1859   Darwin publishes *On the Origin of Species*

1862	**Billroth describes "pyrogen" in injected fluid that causes fever**
1862	**Pasteur publishes the germ theory of disease**
1865	Kekulke deduces the ring structure of the first organic chemical, benzene
1865	**Pasteur begins work on silkworms in Alés, France**
1868	**deBary and Bernhard develop the concept of symbiosis**
1869	Miescher isolates DNA
1870	**Pasteur publishes *Studies of Silkworm Disease***
1873–1889	leprosy, typhoid, pneumonia, tuberculosis, diphtheria, cholera, and tetanus are found to be bacterial
1874	**Panum proposes that fever is caused by gram-negative bacteria**
1876	Koch identifies anthrax bacterium as cause of disease
1879	**deBary coins the term *symbiosis***
1880	*Science* magazine established
1881, 1885	Pasteur develops vaccines for anthrax and rabies
1882	Koch's postulates, how to prove a microorganism causes a disease
1882	**Metchnikoff observes phagocytosis in starfish larva**
1884	Gram develops the Gram stain to distinguish among bacteria
1884	**Koch proves that cholera is caused by a bacterium**
1885	**Cheshire and Cheyne describe foulbrood disease of bees**
1885–1909	**Howell, Loeb, and Blanchard study blood of horseshoe crab**
1886	**Balbini observes defensive reactions to bacteria in insects**
1887	**Blochmann observes bacteria in special organs in roaches**
1887	**Metchnikoff joins Pasteur Institute group**
1888	Pasteur Institute founded
1889	first course in microbiology, taught by Roux at Pasteur Institute
1890, 1906	**Bolle shows that crystals of silkworm jaundice are infectious agents**
1891	Ehrlich proposes that antibodies cause immunity
1892	Dreisch clones sea urchin
1892	Ivanowski discovers tobacco mosaic virus
1898	Beijerinck first uses the term *filterable virus*

Postscript	1900	Reed and others find that yellow fever is a mosquito-borne virus
	1901	**Ishiwata describes *Bacillus thuringiensis* in silkworm**
	early 1900s	phase-contrast and dark-field microscopes invented
	1902	Garrod links inheritance to proteins
	1902	malaria life cycle discovered
	1903	Wright brothers' first flight
	1904	Bateson coins the term *genetics*
	1904	**first vertebrate cell culture**
	1905	first human genetic disease diagnosed (brachydactyly)
	1908	Landsteiner and Popper show that polio is an infectious disease
	1909	**Rhinoceros beetle introduced into Western Samoa**
	1909	**Berliner describes *Bacillus thuringiensis* in flour moths**
	1910	***Entomophaga maimaiga* first imported from Japan for gypsy moth control**
	1910	**Pierantoni and Sulk describe organisms within aphids**
	1911	**Mattes reisolates *Bacillus thuringiensis* from flour moth**
	1911	Rous discovers virus causing cancer
	1913	Morgan and Sturtevant make the first genetic map (fruit fly)
	1913	**Glaser and Chapman show that wilt disease of gypsy moth is a virus**
	1917	**White shows that sacbrood of bees is a virus**
	1919	**Metalnikov joins Pasteur Institute, continues studies on insect immunity**
	1920	**wilt diseases of caterpillars recognized as virus**
	1921	**Buchner publishes *Endosymbiosis of Insects with Plant Microorganisms***
	1924	**Paillot sees replication of caterpillar virus in nucleus and discovers granules**
	1924	**Hertig and Wolbach observe bacteria in gonads of *Culex* mosquitoes**
	1924	Svedberg designs the ultracentrifuge
	1928	Griffith finds "transforming factor" in bacteria (plasmid)
	1929	Fleming finds penicillin
	1929	Levene describes A, T, C, G as building blocks of DNA
	1929	**Glaser finds nematodes in Japanese beetles**
	1931	Ruska and Knoll invent the electron microscope
	1933	**Paillot publishes *Infection of Insects***

1937	Ruska photographs first virus using electron microscope	*181*
1938	**first commercial *Bacillus thuringiensis* product, Sporein, developed in France**	*Postscript*
1941	Waksman develops streptomycin	

1942–1953

1940s–1950s	development of electron microscope for commercial use
1942–1944	**Bergold observes polyhedral viruses**
1944	Avery, MacLeod, and McCarty show that the "transforming factor" is DNA
1944	**Balch and Bird use polyhedral viruses to control spruce sawfly in Canada**
1945	**Steinhaus establishes first U.S. Laboratory of Insect Pathology at the University of California, Berkeley**
1946	Tatum and Lederberg discover conjugation in bacteria
1946	Canadian Laboratory of Insect Pathology is established in Sault Ste Marie, Ontario
1946	**Steinhaus publishes *Insect Microbiology***
1947–1949	Bergold, Hughes, Steinhaus, and Wasser publish electron micrographs of polyhedral virus
1948	World Health Organization formed
1949	Enders, Robbins, and Weller grow polio in cell culture
1949	**Steinhaus publishes *Principles of Insect Pathology***
1950s	***Bacillus thuringiensis* products (Sporein) used in European agriculture**
1950	Hershey and Chase show that DNA is genetic material of viruses
1951	**Toumanoff and Vago conclude that *Bacillus thuringiensis* and *B. cereus* var. *alesti* are the same**
1952	Dulbecco devises virus plaque assay
1952	Chargaff produces "Chargaff's rules" for DNA: A=T, C=G
1952	Wilkins and Franklin use X-ray crystallography to show repeating structure of DNA

1953–1963

1953	Watson and Crick publish DNA double-helix structure
1953	**Hannay sees parasporal bodies in *Bacillus thuringiensis***
1954	**Grace makes first insect cell line, from eucalyptus caterpillar**
1954	**Angus extracts toxin from *Bacillus thuringiensis* crystals**

1956	Bang observes clotting in horseshoe crab blood
1956	World Health Organization initiates Global Malaria Eradication Program
1958	Meselson and Stahl show how DNA replicates
1958–1959	Briggs and Stevens describe inducible immunity in insects
1958	first International Congress of Insect Pathology and Biological Control, Prague
1959	*Journal of Insect* (later *Invertebrate*) *Pathology* established
1960	first aerial spraying of *Bacillus thuringiensis* in New Brunswick, Canada
1960	Kornberg demonstrates DNA synthesis in cell free extract
1960	Jacob and Monod report genetic control of enzyme and virus synthesis
1961	registration of Thuricide, first *Bacillus thuringiensis* product in the United States
1961–1963	triplet universal DNA code for amino acids revealed
1962	Grace reports development of several continuously growing insect cell lines
1962	World Health Organization establishes Tropical Disease Research Working Group
1962	Rachel Carson publishes *Silent Spring*
1962	Hjerten and Mosbach develop column chromatography to purify proteins
1960s	de Barjac and Bonnefoi classify strains by serotyping of flagella; Krywienczyk classifies strains by toxin serotype

1963–1980

1963	Steinhaus publishes *Insect Pathology, an Advanced Treatise*
1963	Huger finds virus in rhinoceros beetle in Malaysia
1964	Levin and Bang find gel formation in limulus lysate
1967	Marschall introduces virus into Western Samoa to control rhinoceros beetle
1967	Society for Invertebrate Pathology founded
1968	Arber, Smith, and Nath use restriction enzymes to cut DNA
1970	Laemmli develops SDS-polyacrylamide gel electrophoresis to separate proteins
1970	Thiemann and Baltimore demonstrate reverse transcription in viruses
1970	Dulmage discovers HD-1 strain of *Bacillus thuringiensis*

1970	Hink modifies insect cell culture medium with fetal bovine serum, other agents
1970	Goodwin and colleagues produce nuclear polyhedrosis viruses in cell culture
1970	Singer isolates *Bacillus sphaericus* SSII-1
1972, 1973	Berg, Cohen, and Boyer make first recombinant DNA
1972–1978	baculovirus genes cloned
1973	first limulus amoebocyte lysate assay approved in United States
1973	Hink and Vail develop plaque assay for nuclear polyhedrosis viruses
1973	Colwell and colleagues find cholera on crustaceans
1974	World Health Organization launches West Africa Onchocerciasis Control Program
1975	Asilomar conference on safety of genetic manipulation
1975	first virus-containing insecticide product for corn earworm approved in United States
1975	Singer isolates *Bacillus sphaericus* strain 1593
1976	Genentec, first biotech company, founded
1976	first virus-containing insecticide product for forest pests approved in United States
1976	Apple introduces the concept of personal computers
1976	Powning and Davidson isolate lysozyme from insect blood
1977	Sanger, Maxam, and Gilbert do first DNA sequencing
1977	Goldberg and Margalit discover *Bacillus thuringiensis israelensis* in Israel
1978	first test-tube baby born
1978	Poinar finds symbiotic bacteria in parasitic nematodes

1980–2005

1980s	sequence of events in baculovirus infections is described
1980	Hultmark and Boman isolate cecropins from insects
1982	Pruisner finds prions
1982	Gonzalez and Carleton find that *Bacillus thuringiensis* genes are on plasmids
1982	Schnepf and Whitely clone first *Bacillus thuringiensis* toxin gene
1983	Montagnier and Gallo announce that HIV causes AIDS
1983	Smith, Fraser, and Summers develop baculovirus expression system
1983	*Bacillus thuringiensis* var. *tenebrionis* discovered

1985	Mullis invents polymerase chain reaction machine used to sequence genes
1985	**first sequences of *Bacillus thuringiensis* toxin genes published**
1986	Hood and Smith automate DNA sequencing
1987	Human Genome Project officially begins in United States
1987	**first HIV vaccine produced using baculovirus expression system**
1989	***Entomophaga maimaiga* reappears in U.S. gypsy moth populations**
1989	**first field trials of recombinant baculoviruses**
1990	BLAST algorithm developed to align gene sequences
1991	World Wide Web initiated
1991	**Li, Carrol, and Ellar report first 3-dimensional structure of *Bacillus thuringiensis* toxin**
1991	**Baumann laboratory describes *Buchnera aphidicola* as symbiont of aphids**
1994	first full sequence of two nuclear polyhedrosis virus genomes
1995	first full gene sequence of free-living organism, bacterium *Haemophilus influenza*
1995	Brown invents DNA microarray "gene chips"
1995	***Bacillus thuringiensis* toxin–containing genetically modified field corn approved for sale in United States**
1996	first large-scale release of genetically modified plants for sale
1997	Dolly the sheep cloned
1999	first full human chromosome sequence published
1999	**experimental malaria vaccine produced in baculovirus expression system**
2000	fruit fly genome sequenced
2001	human genome published in *Science* and *Nature*
2003	**malaria and *Anopheles* genomes published**
2004	***Wolbachia* genome sequenced**
2004–2005	**desert locust outbreak in Africa**

Suggested Readings

1. Pasteur, Silkworms, and the Germ Theory of Disease

Cuny, H. 1966. *Louis Pasteur: The Man and His Theories*. Eriksson, New York.

Dubos, R. 1960. *Pasteur and Modern Science*. Science Tech, Madison, WI.

Holmes, S. J. 1924 (reprinted in 1961). *Louis Pasteur*. Dover, New York.

Steinhaus, E. A. 1975. *Disease in a Minor Chord*. Ohio State University Press, Columbus.

Vallery-Radot, P. 1958. *Louis Pasteur: A Great Life in Brief*. Knopf, New York.

2. Of Caterpillars and Crystals

Aizawa, K. 2001. *Shigetane Ishiwata: His Discovery of Sotto-kin* (Bacillus thuringiensis) *in 1901 and Subsequent Investigations in Japan*. Proceedings of a Centennial Symposium Commemorating Ishiwata's Discovery of *Bacillus thuringiensis*. Kurume, Japan.

Angus, T. A. 1965. Bacterial pathogens of insects as microbial insecticides. *Bacteriological Reviews* 29:364–72.

Beegle, C. C., and T. Yamamoto. 1992. History of *Bacillus thuringiensis* Berliner research and development. *Canadian Entomologist* 124:587–616.

Entwhistle, P. F., J. S. Cory, M. J. Bailey, and S. Higgs, eds. 1993. Bacillus thuringiensis, *an Environmental Biopesticide: Theory and Practice*. J. Wiley and Sons, Hoboken, NJ.

Federici, B. A. 2005. Insecticidal bacteria: An overwhelming success for invertebrate pathology. *Journal of Invertebrate Pathology* 89:30–38.

Lacey, L. A., R. Frutos, H. K. Kaya, and P. Vail. 2001. Insect pathogens as biological control agents: Do they have a future? *Biological Control* 21:2230–48.

Lord, J. C. 2005. From Metchnikoff to Monsanto and beyond: The path of microbial control. *Journal of Invertebrate Pathology* 89:19–29.

3. Out of Africa

Beaty, B. J., and W. C. Marquardt, eds. 1996. *The Biology of Disease Vectors*. University of Colorado Press, Niwot.

Clarke, T. 2004. Malaria is killing one African child every 30 seconds. *Nature News Service*, 25 April.

De Barjac, H., and D. J. Sutherland, eds. 1990. *Bacterial Control of Mosquitoes and Black Flies*. Rutgers University Press, New Brunswick, NJ.

USA Communicable Disease Centers. 2004. West Nile virus. http://www.cdc.gov/ncidod/dvbid/westnile/background.htm.

World Health Organization Tropical Disease Research. 2004. Dengue. http://www.who.int/tdr/diseases/dengue/diseaseinfo.htm.

World Health Organization Tropical Disease Research. 2004. Onchocerciasis. http://www.who.int/tdr/diseases/oncho/diseaseinfo.htm.

4. The Virus That Cures

Arif, B. 2005. A brief journey with insect viruses. *Journal of Invertebrate Pathology* 89:39–45.

Benz, G. A. 1986. Introduction: Historical perspectives. In *The Biology of Baculoviruses*, vol. 2, *Practical Application for Insect Control*, ed. R. R. Granados and B. A. Federici, pp. 1–35. CRC Press, Boca Raton, FL.

Blissard, G. W., and G. F. Rohrmann. 1990. Baculovirus diversity and molecular biology. *Annual Review of Entomology* 35:127–55.

Bonning, B. C., and B. D. Hammock. 1996. Development of recombinant baculoviruses for insect control. *Annual Review of Entomology* 41:191–210.

Doane, C. C., and M. L. McManus, eds. 1981. *The Gypsy Moth: Research toward Integrated Pest Management*. U.S. Forest Service Animal Plant Health Inspection Service Technical Bulletin 1584. U.S. Department of Agriculture, Washington, DC.

Granados, R. R., and B. A. Federici, eds. 1986. *The Biology of Baculoviruses*. 2 volumes. CRC Press, Boca Raton, FL.

Maeda, S. 1989. Expression of foreign genes in insects using baculovirus vectors. *Annual Review of Entomology* 34:351–72.

Martignoni, M. E. 1999. History of TM BioControl-1: The first registered virus-based product for control of a forest insect. *American Entomologist* 45(1):30–37.

Miller, L. K., ed. 1997. *The Baculoviruses*. Plenum Press, New York.

Shuler, M. L., H. A. Wood, R. R. Granados, and D. A. Hammer, eds. 1995. *Baculovirus Expression Systems and Biopesticides*. Wiley-Liss Press, New York.

Steinhaus, E. A. 1949. *Principles of Insect Pathology*. McGraw-Hill, New York.

Steinhaus, E. A., ed. 1963. *Insect Pathology, an Advanced Treatise*, vol. 1. Academic Press, New York.

Vlak, J. M, C. D. de Gooijer, J. Tramper, and H. G. Miltenburger, eds. 1996. Insect cell cultures: Fundamental and applied aspects. *Cytotechnology* 20:173–89.

World Health Organization. 1999. Malaria vaccine progress. *Bulletin of the World Health Organization* 77:361.

5. Scourge of the South Pacific

Bedford, G. O. 1986. Biological control of the rhinoceros beetle in the South Pacific by baculovirus. *Agriculture, Ecosystems, and Environment* 15:141–47.

Huger, A. M. 1966. A virus disease of the Indian rhinoceros beetle *Oryctes rhinoceros*, caused by a new type of insect virus, *Rhabdionvirus oryctes*. *Journal of Invertebrate Pathology* 8:38–51.

Huger, A. M. 2005. The *Oryctes* virus: Its detection, identification, and implementation in biological control of the cocoanut palm rhinoceros beetle, *Oryctes rhinoceros*. *Journal of Invertebrate Pathology* 89:78–84.

Marschall, K. J. 1970. Introduction of a new virus disease of the coconut rhinoceros beetle in Western Samoa. *Nature* 225:288–89.

Young, E. C. 1986. The Rhinoceros Beetle Project: History and review of the research programme. *Agriculture, Ecosystems, and Environment* 15:149–66.

6. Bug against Bug

Boman, H. G. 1995. Peptide antibiotics and their role in innate immunity. *Annual Review of Entomology* 13:61–92.

Boman, H. G., and D. Hultmark. 1987. Cell-free immunity in insects. *Annual Review of Microbiology* 41:103–26.

Brey, P. T., and D. Hultmark. 1998. *Molecular Mechanisms of Immune Responses in Insects*. Chapman and Hall, London.

7. Saved by a Crab

Ecological Research and Development Group. 2004. The horseshoe crab. http://www.horseshoecrab.org/con/con.html.

Garland, M. S. 2004. An ancient lineage worth saving. *Science* 304:1113.

Gupta, A. P., ed. 1990. *Immunology of Insects and Other Arthropods*. CRC Press.

Holloway, M. 2002. Ancient rituals on the Atlantic Coast. *Scientific American* (Mar.):94–97.

University of Delaware Graduate College of Marine Studies. 2004. LAL research. http://www.ocean.udel.edu/horseshoecrab/research/lal.html.

8. Sea Sickness

Couch, J. A., and J. W. Fournie, eds. 1993. *Pathobiology of Marine and Estuarine Organisms*. CRC Press, Boca Raton, FL.

Davidson, E. W., ed. 1981. *Pathogenesis of Invertebrate Microbial Diseases*. Allanheld, Osmun, Totowa, NJ.

Ewart, J. W., and S. E. Ford. 1993. History and impact of MSX and Dermo diseases on oyster stocks in the northeast region. *Northeastern Regional Aquaculture Center Fact Sheet no. 200*. North Dartmouth, MA.

Lee, R. F., and M. E. Frischer. 2004. The decline of the blue crab. *American Scientist* 92:548–53.

Lightner, D. V. 1999. The penaeid shrimp viruses TSV, IHHNV, WSSV, and YHV: Current status in the Americas, available diagnostic methods, and management strategies. *Journal of Applied Aquaculture* 9:27–51.

NOAA Research. 2005. Battling oyster disease in the Chesapeake. http://www.oar.noaa.gov/spotlite/archive/spot_oysters.html.

Queensland Government Department of Primary Industries and Fisheries. Oyster culture in Queensland. 2005. http://www.dpi.qld.gov.au/fishweb/11222.html.

Shrimp News International. 2005. About shrimp farming. http://www.shrimpnews.com/about.html.

Sparks, A. K. 1985. *Synopsis of Invertebrate Pathology Exclusive of Insects*. Elsevier, New York.

Stewart, J. E. 1993. *Pathobiology of Marine and Estuarine Organisms*. CRC Press, Boca Raton, FL.

9. Deadly Hitchhikers

Colwell, R. R., A. Huq, M. S. Islam, K. M. A. Aziz, M. Yunus, N. H. Khan, A. Mahmud, R. B. Sack, G. B. Nair, J. Chakraborty, D. A. Sack, and E. Russek-Cohen. 2003. Reduction of cholera in Bangladeshi villages by simple filtration. *Proceedings of the U.S. National Academy of Sciences* 100:1051–55.

Faruque, S. M., D. A. Sack, R. B. Sack, R. R. Colwell, Y. Takeda, and G. B. Nair. 2003. Emergence and evolution of *Vibrio cholerae* 0139. *Proceedings of the U.S. National Academy of Sciences* 100:1304–09.

Huq, A., B. Xu, M. A. R. Chowdhury, M. S. Islam, R. Montilla, and R. R. Colwell. 1996. A simple filtration method to remove plankton-associated *Vibrio cholerae* in raw water supplies in developing countries. *Applied and Environmental Microbiology* 62:2508–12.

Pollitzer, R. 1959. *Cholera*. World Health Organization, Geneva.

Wachsmuth, I. K., P. A. Blake, and O. Olsvik, eds. 1994. Vibrio cholerae *and Cholera: Molecular to Global Perspectives*. American Society for Microbiology Press, Washington, DC.

10. Cloak and Dagger

Kaya, H. K., and R. Gaugler. 1993. Entomopathogenic nematodes. *Annual Review of Entomology* 38:181–206.

Liu, J., G. O. Poinar Jr., and R. E. Berry. 2000. Control of insect pests with entomopathogenic nematodes: The impact of molecular biology and phylogenetic reconstruction. *Annual Review of Entomology* 45:287–306.

Ohio Agricultural Research and Development Center. 2003. Biology and ecology of insect parasitic nematodes. http://www.oardc.ohio-state.edu/nematodes/biologyecology.htm.

Stock, S. P. 2005. Insect parasitic nematodes: From lab curiosities to model organisms. *Journal of Invertebrate Pathology* 89:57–66.

11. Bad News for the Good Guys

Bailey, L., and B. V. Ball. 1991. *Honey Bee Pathology.* 2nd ed. Academic Press, New York.

Berenbaum, M. R. 1995. *Bugs in the System: Insects and Their Impact on Human Affairs.* Addison-Wesley, Reading, MA.

Morse, R. A., and R. Nowogrodzki, eds. 1990. *Honey Bee Pests, Predators, and Diseases.* 2nd ed. Cornell University Press, Ithaca, NY.

Needham, G. R., R. E. Page Jr., M. Delfinado-Baker, and C. E. Bowman, eds. 1988. *Africanized Honey Bees and Bee Mites.* Wiley, New York.

Winston, M. L. 1987. *The Biology of the Honey Bee.* Harvard University Press, Cambridge, MA.

12. The Mysterious Reappearing Fungus

Doane, C. C., and M. L. McManus, eds. 1981. *The Gypsy Moth: Research toward Integrated Pest Management.* U.S. Forest Service Animal Plant Health Inspection Service Technical Bulletin 1584. U.S. Department of Agriculture, Washington, DC.

Hajek, A. E. 1999. Pathology and epizootiology of *Entomophaga maimaiga* infections in forest Lepidoptera. *Microbiology and Molecular Biology Reviews* (Dec.): 814–35.

Howard, L. O. 1930. *A History of Applied Entomology (Somewhat Anecdotal).* Smithsonian Miscellaneous Collections 84. Smithsonian Institution, Washington, DC.

Reardon, R. C., and A. E. Hajek. 1998. The gypsy moth fungus *Entomophaga maimaiga* in North America. USDA Forest Service Technology Transfer Bulletin FHTET-97-11. U.S. Department of Agriculture, Washington, DC.

Weseloh, R. M. 2003. People and the gypsy moth: A story of human interactions with an invasive species. *American Entomologist* 49:180–90.

13. The Sixth Plague of Egypt

Enserink, M. 2004. Can the war on locusts be won? *Science* 306:1880–82.

Food and Agriculture Organization. 2004. Desert locust update, 17 December. http://www.fao.org/news/global/locusts/locuhome.htm.

Goettel, M. S., and D. L. Johnson. 1997. *Microbial Control of Grasshoppers and Locusts.* Memoirs of the Entomological Society of Canada 171. Ottawa, Ontario.

Limson, J. 2004. Green Muscle: An environmentally friendly weapon against locust and grasshopper plagues. *Science in Africa*, issue 1 (31 Dec.). http://www.scienceinafrica.co.za/Green_Muscle.htm.

Lomer, C. J., R. P. Bateman, D. L. Johnson, J. Langewald, and M. Thomas. 2001. Biological control of locusts and grasshoppers. *Annual Review of Entomology* 46:667–702.

LUBILOSA. 1999. Green Muscle user handbook. http://www.lubilosa.org/dwnload.htm.

Neuenschwander, P. 2004. Harnessing nature in Africa. *Nature* 432:801–2.

Organization for Economic Co-operation and Development. 2004. Desert locust outbreaks in West Africa. http://oecd.org.

Pearson, H. 2004. Africa's locust crisis worsens. *News@Nature.com* (20 Aug.).

UN Development Programme Equator Initiative. 2005. Green Muscle–Benin. http://www.tve.org.

14. Ad Infinitum

Batra, L. R., ed. 1997. *Insect-Fungus Symbiosis: Nutrition, Mutualism, and Commensalisms*. John Wiley, Hoboken, NJ.

Baumann, P., L. Baumann, C.-Y. Lai, D. Rouhbakhsh, N. Moran, and M. A. Clark. 1995. Genetics, physiology, and evolutionary relationships of the genus *Buchnera*: Intracellular symbionts of aphids. *Annual Review of Microbiology* 49:55–94.

Baumann, P., M. A. Munson, C.-Y. Lai, M. Clark, L. Baumann, N. Moran, and B. C. Campbell. 1993. Origin and properties of bacterial endosymbionts of aphids, whiteflies, and mealybugs. *American Society for Microbiology News* 59:21–24.

Beckage, N. E., and D. B. Gelman. 2004. Wasp parasitoid disruption of host development: Implications for new biologically based strategies for insect control. *Annual Review of Entomology* 49:299–330.

Bourtzis, K., and T. A. Miller, eds. 2003. *Insect Symbiosis*. CRC Press, Boca Raton, FL.

Douglas, A. E. 1994. *Symbiotic Interactions*. Oxford University Press, Oxford, UK.

O'Neil, S. L., A. A. Hoffmann, and J. H. Werren, eds. 1997. *Influential Passengers: Inherited Microorganisms and Arthropod Reproduction*. Oxford University Press, Oxford, UK.

Pew Initiative on Food and Biotechnology. 2004. http://pewagbiotech.org.

Tanada, Y., and H. K. Kaya. 1993. *Insect Pathology*. Academic Press, New York.

Werren, J. H. 1997. Biology of *Wolbachia*. *Annual Review of Entomology* 42:587–609.

Illustration Credits

Agostino Maria Bassi of Lodi, Italy. From J. R. Porter, *Bact. Reviews* 37(1973): 284–85. Reprinted by permission of the American Society for Microbiology. Photo courtesy of Donald Roberts.

Pasteur and wife at Alés. From Institut Pasteur archives, Paris, image number D3072. Reproduced by permission, approval number 03295.

Shigetane Ishiwata. From B. Federici, *J. Invertebr. Pathol.* 89(2005):30–31. Reprinted by permission of Elsevier, Inc. Photo courtesy of Keio Aizawa.

Edward Steinhaus, John Briggs, Constantin Toumanoff. Photographer unknown.

Light micrograph of *Bacillus thuringiensis*. Reproduced by permission of J. Weiser.

Onchocerciasis victim. Photo courtesy of the World Health Organization/ Tropical Disease Research Photo Collection, image number 01031979. Reproduced by permission.

Anopheles mosquito biting, exuding fluids. Reproduced by permission of C. Berry.

Electron micrograph of *Bacillus sphaericus* spore and toxin crystal. Photo by E. Davidson.

Joel Margalit collecting soil samples in the Negev Desert. Photo by E. Davidson, reproduced with permission of J. Margalit.

Electron micrograph of nuclear polyhedrosis virus. Reproduced by permission of B. Federici.

Caterpillar killed by nuclear polyhedrosis virus. Photo by the late C. G. Thompson.

Max Summers. Photo by John Briggs, with permission of M. Summers.

Rhinoceros beetle. Photo courtesy of A. Huger, from *J. Invert. Pathol.* 89(2005): 80. Reproduced by permission of Elsevier, Inc.

Healthy coconut palm and palm damaged by *Oryctes* rhinoceros beetle. Courtesy of A. Huger.

Elie Metchnikoff. From Institut Pasteur archives, Paris, image number D380. Reproduced by permission, approval number 03295.

Beetle grub infested by nematode worms. Photo by R. Gaugler, reprinted by permission of CABI Publishing and R. Gaugler.

Langstroth moveable-frame beehive. Photo by G. Reuter, courtesy of Marla Spivak.

Electron micrograph of *Paenibacillus larvae* penetrating midgut cells of a bee larva. Photo by E. Davidson.

Entomophaga maimaiga spores. From A. Hajek, *Microbiol. Molec. Biol. Rev.* 63(1999):818. Reprinted by permission of the American Society for Microbiology and A. Hajek.

Gypsy moth caterpillar killed by *Entomophaga*. Courtesy of A. Hajek.

Grasshopper infected with *Metarhizium anisopliae*. Courtesy of M. Goettel.

Symbiotic bacteria in the gut of a louse. Drawing by G. Buchner, from G. Buchner, *Endosymbiose der Tier mit Pflanzlichen Mikroorganismen* (Springer Verlag, Heidelberg, Germany, 1965), 485. Reprinted by permission of Springer Science and Business Media.

Index

About the Author

Elizabeth West Davidson spent her childhood collecting insects in the fields and woods near the small town of Damascus, Ohio. When she read an old copy of Edwin Way Teal's *Fascinating Insect World of J. Henri Fabre*, she came to realize that people can actually spend their lives studying insects. These experiences, along with the encouragement of her professors at Mount Union College, led her to enter the graduate program in entomology at Ohio State University. She planned to do research in toxicology, but accidental contamination of her research beetles with the fungus *Beauveria bassiana* directed her research instead to insect pathology. She received a Ph.D. in entomology from Ohio State University in 1971. Since 1973 she has been a member of the research faculty at Arizona State University in Tempe, where she has worked on development of bacterial control agents for mosquitoes and the silverleaf whitefly, a serious pest of cotton. Since 1995 she has also studied diseases associated with worldwide declines of amphibian populations. Davidson is author of more than one hundred peer-reviewed publications, has edited or co-edited three books, and has served as president of the Society for Invertebrate Pathology.